T0313015

Power Transmission and Distribution

2nd Edition

Power Transmission and Distribution

2nd Edition

Anthony J. Pansini, E.E., P.E.

Routledge
Taylor & Francis Group
LONDON AND NEW YORK

Published 2020 by River Publishers

River Publishers

Alsbjergvej 10, 9260 Gistrup, Denmark

www.riverpublishers.com

Distributed exclusively by Routledge

4 Park Square, Milton Park, Abingdon, Oxon OX14 4RN

605 Third Avenue, New York, NY 10158

First published in paperback 2024

Library of Congress Cataloging-in-Publication Data

Pansini, Anthony J.
    Power transmission and distribution/Anthony J. Pansini.--2nd ed.
        p. cm.
    Includes index.
    ISBN: 978-0-8493-5034-4 (print) — 978-8-7702-2249-5 (electronic)
        1. Electric power transmission. 2. Electric power distribution. I.
Title.

    TK3001.P29 2005
    621.319--dc22

                                                        2004056439

Power transmission and distribution, second edition/Anthony J. Pansini
First published by Fairmont Press in 2005.

Routledge is an imprint of the Taylor & Francis Group, an informa business

Publisher's Note
The publisher has gone to great lengths to ensure the quality of this reprint but points out that some imperfections in the original copies may be apparent.

978-0-88173-503-1 (The Fairmont Press, Inc.)
978-8-7702-2249-5 (online)

ISBN: 978-0-8493-5034-4 (hbk)
ISBN: 978-87-7004-588-9 (pbk)
ISBN: 978-0-203-74264-8 (ebk)
ISBN: 978-1-003-15120-3 (eBook+)

*In memory of my parents
in appreciation of their sacrifices
and encouragement*

Contents

Preface to the First Edition

It has been some time since a book was written on power transmission and distribution, a book that can be used as a textbook for the many for whom this subject, for one reason or another, may be of interest. In one place, there can be found the electrical, mechanical and economic considerations associated with the successful planning, design, construction, maintenance and operation of such electrical systems.

Simple explanations of materials and equipment describe their roles in the delivery of power, in small and large quantities, to homes and offices, farms and factories. They meet the needs of nontechnical people, including the legal and financial sectors, as well as those whose interests may involve the promotion of equipment sales and maintenance, public information, governmental and other functions and activities. For the neophyte engineer and the seasoned operator, the practical technical discussion provides reference and review of the bases and tools employed in meeting the problems that arise in their daily endeavors. And, finally, the student and researcher will find sufficient theory and mathematical analyses to satisfy their thirst for knowledge and to impress their neighbors with the depth of their intellect!

Both the young who enjoy the benefits of modern electrical supply and the older groups who have seen and experienced the remarkable development made in its transmission and distribution must recognize that such advances are the work of many to whom a debt is due. And to some of us who have been given the privilege of making even slight contributions, we are grateful for the opportunities afforded us during a most enjoyable and fulfilling career.

Thanks are extended to the people who have been helpful along the way, too many to name individually, and to the staff of The Fairmont Press who have aided in the preparation and publishing of this work. The contributions of material and illustrations by the manufacturers for which I am extremely grateful, are especially acknowledged. In any work, errors somehow manage to intrude, and for any of these, I take sole responsibility. Finally, a deep acknowledgment to my beloved wife for her unstinted support, patience and understanding through the many years in which I have been engaged in this and kindred endeavors.

Anthony J. Pansini
Waco, Texas
1990

Preface to the Second Edition

Some twenty years have passed since the original publication of this book, normally sufficient to warrant an updating dictated by events and heralding the arrival of a new century. The explosion of electronically operated devices (computers, robots, automatic controls, etc.) have required micro refinements in the quality of electric supply that could not tolerate those associated with the macro commercial supply of this commodity; necessary *corrective actions* peculiar to each such application were (and are) undertaken by the individual consumer. But the continually increasing dependence on electricity in practically every one of life's endeavors also called for improvements in the quality standards of its supply to which this updating is addressed.

Notable events during this twenty-year period that helped in calling for better quality standards for those elements associated with reliability include the deregulation of electric (and other) utilities, the events of September 11, 2001, and the blackouts on northeast North America on August 14, 2004, in the London area and Italian peninsula within two weeks of each other. And on the positive side, the proliferation of automation brought about by the blooming electronic technology.

Transmission systems have been the subject of the greater changes. Under deregulation, their role in the supply chain has been essentially reversed, from being the back up and peak supplier in generation-based systems, they become the main source of supply with generation reduced to a minimum if not entirely eliminated (to reduce capitalization and its effect on rate structures in a competitive market) Figure P-1. For economic and environmental reasons, transmission lines are situated in areas of sparse population making them subject to the vagaries of man and nature, tailor made for assaults by vandals and saboteurs. Finally, with transmission lines connected together in a grid, supposedly for better reliability and economy, failures causing the outage of a line may cause another of the lines to trip open from overload, causing another and another line to "cascade" open until total area blackout occurs.

It appears, quite unexpectedly, that the application of loop circuits substantially improves the reliability of such transmission lines. Loop

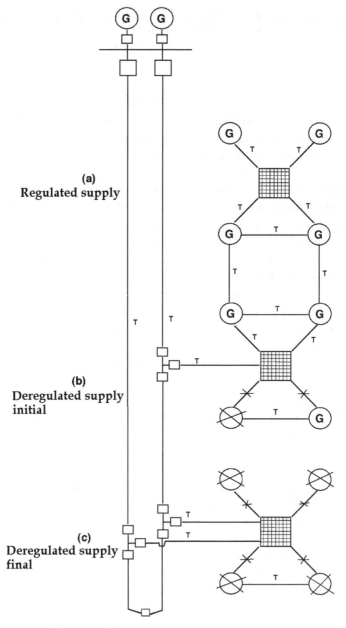

Figure P-1. Simplified schematic diagram of transition from regulated to deregulated supply systems.

circuits essentially provide a two way feed to the consumers, insuring them continuity of service should a fault develop on the circuit (except for those situated on the section on which the fault occurs) and especially if both halves of the loop circuit are not mounted on the same supporting structures. The reliability of the deregulated line is enhanced, and similarly, the damage inflicted by a saboteur or vandal may be limited to a section of the line. In the case of the transmission grid, supplanting it with a number of loop circuits not only removes the possibility of lines cascading open from overloads or instability, but permits the circuits to be loaded nearer their full capacity.

Distribution systems have also been affected by these events, although not in the same manner of vulnerability as transmission systems. Where additional generation, and/or transmission was not available, or too great an expenditure to supply some additional distribution loads, distributed generation made its entry on the scene. Here small generating units, usually powered by small gas turbines, are connected directly to the distribution system, in the same manner as larger cogeneration units. These units may be both consumer- or utility-owned and operated, and may constitute safety hazards.

The chapter on street lighting is relegated to the appendices not only as essentially obsolete, but as an example of constant current circuitry. In its stead is a description of direct-current transmission line with its positive and negative features, but an excellent future feature in the electric supply scenario.

A Texas size thank you goes out to all who have directly or indirectly contributed in the publishing of this work and especially to The Fairmont Press for their help and support.

Chapter 1

Introduction, Consumer Characteristics

INTRODUCTION

The system of delivery of electricity to consumers parallels that of most other commodities. From the generating or manufacturing plant, this product is usually delivered in wholesale quantities or via transmission facilities to transmission substations that may be compared to regional warehouses. From there, the products may (or may not necessarily) be further shipped to jobbers over subtransmission lines to distribution substations or local depots. The final journey delivers these products to retailers via distribution systems that supply individual consumers. One important difference in this comparison is the lack of storage capability (for practical purposes) of electricity; every unit of electricity consumed at any moment must be generated at that same moment. A diagram of an electric utility system indicating the division of operations is shown in Figure 1-1. This work concerns the transmission and distribution elements.

Just as many of the larger manufacturing companies began as small enterprises, so, too, did many of the electric utilities. The first commercial electric system was constructed and placed in operation in 1882 by Thomas Alva Edison in New York City. It was a direct current system that served a limited number of consumers in the vicinity of the plant at a nominal voltage of 100 volts. A number of other small systems (also direct current) in urban and suburban areas were supplied from the generating facilities of manufacturing factories. From these maverick systems that, in some instances, grew like Topsy without planning, the large utility systems were to be formed. Interestingly, almost a century later, privately owned generating facilities of industrial and commercial companies were once again to exercise that same function. The sale of their excess energy through electrical connections to utility companies' transmission and distribution systems is referred to as "cogeneration," and will be discussed more fully later in this work.

The invention of the transformer in 1883 in England by John Gibbs and Lucien Gaullard, together with the invention of the alternating current induction motor and the development of polyphase circuitry in

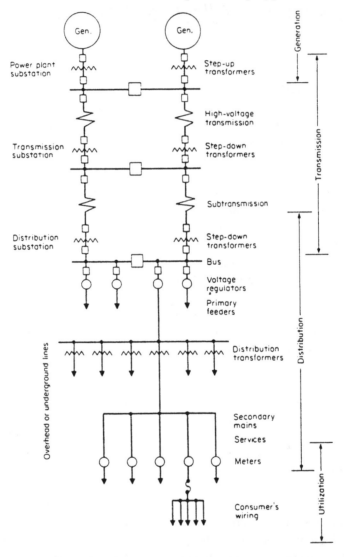

Figure 1-1. Electric System Divisions—Note Overlap
(Courtesy Westinghouse Electric Co.)

1886 by Professor Galileo Ferraris in Turin, opened the way for the adoption of alternating current and the rapid expansion of electrical transmission and distribution systems. Transmission of electric power dates from 1886 when a line was built at Cerchi, near the city of Turin in Italy, to transmit some 100 kilowatts 30 kilometers, employing transformers, to raise and lower a 100 volt source to 2000 volts and back to 100 volts for utilization. In the same year, the first alternating current distribution system in the United States, also using transformers, was put into operation at Great Barrington in Massachusetts: the 100 volt system included two 50 light and four 25 light transformers serving 13 stores, 2 hotels, 3 doctors' offices, a barber shop, telephone exchange and post office from a 500 volt source.

The adoption of alternating current, employing transformers, together with the general public's acceptance of the less than pleasing overhead facilities almost entirely accounts for the unparalleled expansion experienced by the electric industry. The successful combination of transformer and overhead installations exemplifies a basic solution of the electrical, mechanical and economic problems associated with the design of transmission and distribution systems, as well as their construction, maintenance and operation. These three problems, although subject to independent solutions, interact upon each other.

Electrical design considerations are based generally on acceptable values of loss in electrical pressure or voltage drop and those of energy loss. These considerations may be modified to accommodate desired protection, environmental and other requirements. The permissible values determine the size of conductors and the associated insulation requirements. The physical characteristics of the conductors impact on the mechanical designs of such systems.

Mechanical design involves the study of structures and equipment. It includes the selection of proper materials and their combination into structures and systems in such a manner as to meet the electrical design requirements, giving due consideration to matters of strength, safety, temperature variations, length of life, appearance, maintenance, and other related factors.

Economic design includes the investigation of relative costs of two or more possible solutions to the combined electrical and mechanical requirements. The choice is governed (although not necessarily) not by the lowest annual carrying charge on the investment in the systems studied, but by that which is equal (or closest) to the annual cost of losses associated with one of the systems under study. This relationship is known as Kelvin's Law. Many factors intrude, however, to modify the applicable conditions. These factors pertain generally to safety and environmental requirements as well as provision for possible future demands for electric power, creating changes that may affect the several components involved in the solution to design problems; for example, new technology, revised codes and standards, inflation, new reliability and environmental requirements, etc. The final decision must also satisfy the electrical and mechanical design requirements. These criteria apply to both the transmission and distribution systems.

Referring to the diagram in Figure 1-1, although it has been customary to consider generation, transmission and distribution as three interdependent elements constituting a single enterprise, as one electric utility system, financial and conservation considerations have given rise to consideration of each of the three as separate and distinct enterprises. Acquisition of each of the three by independent parties could be a means of diversifying their investments. Problems of cooperation in the operation of such separately owned systems would affect the consumer, and could possibly cause the construction of duplicate competitive systems.

In the presentation of the material that follows, it will be assumed that the reader is familiar with the essentials of electricity, including vector representation, concerning the properties of both direct and alternating current circuits, including resistance, inductance, capacitance, impedance, and their Ohm's Law relationships.

Although the usual flow of electricity to the consumer is from the generating plant through the transmission system into the distribution system, the discussion will treat the delivery system in the reverse order: starting with the consumer and working toward the central generating plant.

CONSUMER CHARACTERISTICS

To begin the electrical design of transmission and distribution systems, it is necessary to know the characteristics of the building blocks upon which the design of the systems is predicated; that is, the consumer to be served. Obviously, each consumer cannot be considered independently, but they may be studied as a class and as groups as they affect the final design of the systems.

For convenience, consumers may be broadly classified as residential, commercial, and industrial. The requirements of each type to be determined include:

1.   The total consumption of electricity over a period of time, (say) annually.

2.   The changes in rate of consumption, (say) hourly, over periods of time: daily, weekly, monthly, annually.

3.   The voltage required for the proper operation of the loads to be served; the tolerance permitted in the variation of this voltage, and whether the rapidity of such variations would cause flicker of lights to result.

4.   The reliability requirements of the loads to be served, that is, the degree of interruption to service, as well as variations in the three items above, that may be tolerated or permitted.

Electric systems consist essentially of conductors in the form of wire, terminals, blades of switches or circuit breakers, wires in transformers, motors, and other equipment. The criteria on which their designs are based are two:

1.   The permissible drop in voltage or pressure of the electricity flowing through them, and

2.   The permissible energy loss caused by electricity flowing through them, manifested in the form of heat to be dissipated harmlessly.

From Ohm's Law, the loss in voltage is equal to the product of the current flowing through a conductor and its resistance:

$$I\,(\text{current}) = \frac{E\,(\text{voltage})}{R\,(\text{resistance})} \text{ from which, } E = IR$$

Energy loss is the product of power and time; power, however, is the product of the voltage imposed on the conductor and the current flowing through it. Again, from Ohm's Law, this can be derived into the product of the square of the current flowing through the conductor and its resistance:

$$P(\text{Power}) \text{ in watts} = E \times I; \text{ or } IR \times I = I^2 R$$

and energy = Power × Time, in watt-hours or kilowatt-hours.

The heat generated must be dissipated if temperature rise is to be limited to safe values (i.e., before failure results, usually in the insulation surrounding a conductor). Also, the heat generated represents a loss of energy for which there is no economic return, and some reasonable value must be placed on its limits. While standards (and guarantees) usually specify a definite temperature limit, e.g., 50°C, 70°C, etc., these figures are not rigid as temperatures (and designs) are affected by ambient temperatures, duration of high temperature, including those preceding the imposition of the condition causing the undesirably high temperature, effect of wind and other cooling factors, etc. These conditions affect the selection of conductors, transformers, switches, and other facilities comprising the transmission and distribution systems.

The consumer's connected load, therefore, becomes the starting point for the design of such systems. An examination of "typical" consumer's connected loads will quickly determine the voltage requirements: 120 volts for lighting and many of the appliances and 240 volts for some of the larger size units; for some large motor loads, polyphase (usually three phase) voltages of 120/208 volts, 120/240 volts, or 277/460 volts. This will determine the number of conductors to supply the consumer's load: 2, 3 or 4 wires, as well as the value of the associated insulation.

The size of the conductor is determined by the highest value of current to be carried. Common sense indicates the consumer will, at various times throughout the day, be using different combinations of the units comprising his connected load. The magnitude of the total current to be supplied over the conductors will, therefore, also vary throughout the day. For design purposes, the maximum current is determined by taking the maximum consumption of electricity over a definite period of time (usually 15, 30 or 60 minutes) and converting it into current or amperes, Figure 1-2.

This value may be different for each day of the week, month or year; hence, the largest of these is taken as the basis for design and is known as the "maximum demand" of the individual consumer. This factor affects the selection of conductors, transformers and other facilities comprising the distribution system.

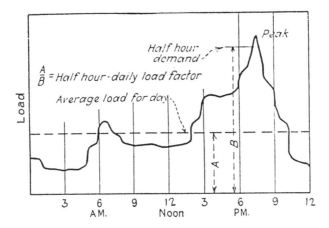

Figure 1-2. Load Factor; Maximum Demand

The distinction between the consumer's demand and connected load is most important. Connected load is the total of the rated capacities of all electric appliances, lights, motors, etc., that are connected to the wiring of a consumer. The actual demand is almost always considerably less than the connected load, because the different units are used at different times, or, if used at the same time, their peak loads may not be simultaneous, or in either case, all units may not be loaded to full capacity at their peak loads. The exception to this is on loads where all utilization equipment is of the same general type and is used at the same time and at the same capacity, such as may be found in some manufacturing plants, in water or sewer treatment plants, or in street lighting circuits. The ratio of maximum power demand to total connected load is called the "demand factor."

The method of determining the demand factor is also applied to the loads of a group of consumers. Here, the combined maximum demand of the group is compared to the total of the maximum demands of each of the members of the group, and this ratio is known as the diversity factor, Figure 1-3. For example, a transformer may supply six consumers whose individual maximum demands total 150 kVA, but whose combined maximum demand may be only 75 kVA. The diversity factor is 150/75 or 2. It should be noted that the demand factor is defined in such a way that it is always less than 1, while diversity factor in such a way it is always greater than 1.

Such diversity exists between consumers, between transformers, between feeders, between substations, etc. It is used in reducing the required capacity of facilities that would otherwise be required if based on connected loads or the sum of component load demands only.

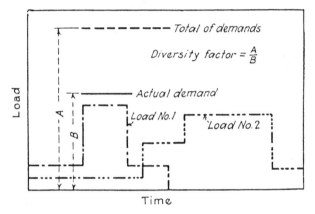

Figure 1-3. Diversity Factor

Load Factor

Demand factors and diversity factors, while basic to the design of distribution circuits, do not include another important element: that of the use made of the facilities installed, the relationship of consumption (which is a measure of revenue and return on investment) to the maximum demand. The consumption can be converted to an average demand by dividing the kilowatt-hours over a stipulated period of time (day, week, month, year) by the time. The ratio of this average demand to the maximum demand is known as the load factor (Figure 1-2) and is an index of the efficiency with which the system or portion of the system under consideration is utilized; 100 percent load factor or 24 hours per day at peak load being the maximum possible.

Loss Factor

A companion factor, the ratio of average power loss for a stipulated period of time (day, week, month, year) to the maximum loss or loss at peak (15, 30, 60 minutes) during the same period. The distinction between the load factor and loss factor is that the former pertains to loads (maximum and average) while the latter pertains to losses which are proportional to the square of the corresponding loads, Figure 1-4.

Equivalent Hours

Associated with the loss factor is a quantity called "equivalent hours." It is defined as the average number of hours per day which the peak load would have to continue to give the same total energy loss as that given by the variable load (throughout the week, month, year, as the case may be). Equivalent hours = loss factor × 24. This factor is useful in determining the cost of energy losses which, in turn, may result in the installation of more economical larger facilities.

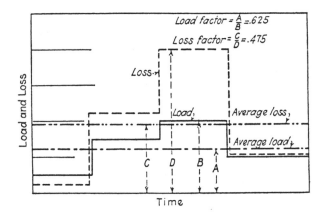

Figure 1-4. Load; Loss; Load Factor; Loss Factor

Another factor, the ratio of the average demand to the installed capacity is called the use factor, and is an indication of how much of the investment is used. It is sometimes used in place of the load factor as an index of the efficiency with which the system under consideration is utilized.

Power Factor

In alternating current circuits, almost always current and voltage will be found out of phase with each other; this relation, together with instantaneous power, and vector representation of these quantities, are shown in Figure 1-5. When loads are designated in kilowatts, it is essential also to know the power factor as the capacities of transformers, capacitors, etc., whose ratings are in kilovolt-amperes. Also, line losses are proportional to the square of the current, and voltage drop proportional to the current.

Power factor, then, is the ratio of power (watts or kilowatts) to the product of voltage and current (in volt-amperes or kilovolt-amperes). It is sometimes defined as the ratio of real power to apparent power. From Figure 1-5

$$\text{Power factor} = \frac{\text{watts}}{\text{EI}} = \cos \theta$$

where E = effective voltage, I = effective current, and θ is their angular displacement in phase.

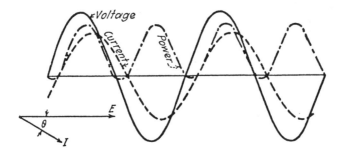

Figure 1 -5. Power Factor

A power factor approaching unity or 100 percent as nearly as possible, is important in the design of the distribution (and transmission) system which is dependent on current capacity. For the same current and the same voltage, the power delivered is directly proportional to the power factor.

Balance

On three wire, single phase, or direct current, 120/240 volt circuits, unbalance often occurs between the loads on the two sides of the circuit, resulting in unbalanced voltages.

Where polyphase (three phase) circuits are employed, usually for large consumers, loads on each of the phases are likely to be unequal. Unequal or unbalanced currents produce unequal voltage drops in lines, transformers, etc., producing unbalanced voltages at the loads that, in turn, produce unbalanced currents in polyphase equipment, e.g., motors. The unbalance may be expressed as a percentage, or balance factor, from a nominal base, or from the average of all of the phase voltages. While phase relations are not indicated, this factor serves as a convenient measure of unbalance.

Coincidence or Diversity Factor

The ratio of the maximum demand of the whole to the sum of the maximum demands of each of the individual consumers is known as the coincidence or diversity factor. (Figure 1-3)

Chapter 2

Distribution System Electrical Design

The design of the distribution system starts with the service to the consumer. A single individual consumer may be supplied, through a transformer, from a higher voltage source, known as the primary. This particular arrangement is employed primarily in rural areas, supplying electricity to farms that are situated remote from each other. It may also be employed in supplying electricity to larger consumers in other areas.

Several consumers may be supplied from one transformer and may have their services terminate on the same pole or structure on which the transformer is located, or the services may be connected to a "secondary main" which is supplied from the transformer, Figure 2-1. In urban and suburban areas, individual consumers may be situated close together in groups. Here, economy may be realized by supplying a number of these consumers from one transformer, as described above. The load imposed on this transformer is not the sum of the maximum demands of each of the consumers connected to it, but a "new" maximum demand of the whole; this is because the maximum demands of each of the individual consumers do not occur at the same time.

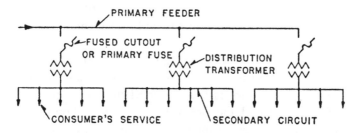

Figure 2-1. Typical Method of Connecting Distribution Transformers Between Feeders and Radial Secondary Circuits. (*Courtesy Westinghouse Electric Co.*)

The single individual consumer supplied through a transformer may be considered as a concentrated load on the distribution system. A group of consumers served from a secondary main may be considered as an essentially uniformly distributed load for design purposes; in most cases the error introduced is negligible. This simplifies the problems of determining voltage drop and power loss over a line on which the load is uniformly distributed. Voltage drop from the source point to the extreme end of such a line is equivalent to that as if the total load is concentrated at one-half the distance. The power loss would be equivalent to that as if the total load is concentrated at one-third the distance.

The solution to the problems indicated above assume that the size of conductor in the secondary main is the same along its entirety. Theoretically, the size of the conductor from the source point to the several points of service connections may be reduced in size toward the end of the main. Practical problems of purchasing, stocking and handling, of connecting conductors of several sizes, etc., make it economically desirable to have one size conductor. Moreover, load growth is often accommodated by dividing the secondary main and installing an additional transformer.

In some instances where greater service reliability is desired, the secondary mains of several adjacent transformers may be connected together (usually through a fuse or other protective device); this is referred to as secondary banking, with all the transformers supplied by the same primary feeder, Figure 2-2. In the event of a failure of a transformer, its load is carried by adjacent transformers, with perhaps some reduction in voltage, but without interruption of service. Other advantages claimed for secondary banking are a better distribution of load among the various transformers, and better average voltage conditions resulting from such load distribution; some advantage is also taken of the diversity between demands on adjacent transformers in reducing the total transformer load.

A disadvantage of secondary banking is the possibility of cascading; that is, if one transformer fails and adjacent transformers become overloaded in picking up its load, beyond their fuse capacity, or fail because of excessive load, the increased load is passed on to other transformers in the bank, causing them also to go out, causing interruption to the whole banked secondary. This occurrence may be minimized by proper fusing between the secondary mains constituting the bank and by proper sizing of transformers in relation to adjacent ones.

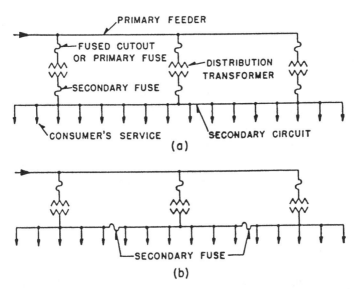

Figure 2-2. Typical Methods of Banking Transformers Supplied by the Same Radial Primary Feeder. *(Courtesy Westinghouse Electric Co.)*

In designing a secondary bank, it is preferable for it to be in the form of a grid, rather than long single mains. It should be so arranged that, if possible, the load dropped by a transformer is fed directly from at least two transformers; the sizes of the transformers and associated fusing should be such as to prevent cascading.

The division of the load depends on the impedance of the various paths through the secondary grid by which the load may be fed, including the impedance of the transformers. These impedances will vary with the size of the transformers, the distances between them, and the size and spacing of the secondary mains. The arrangement of the various individual loads among the secondary and the normal load on each of the transformers will also have an effect on how the load will be supplied.

On overhead lines, the impedance of the secondary mains between transformers is likely to be high compared with that of the transformers, so that a large proportion of the load of a faulty transformer will probably be picked up by the immediately adjacent one. On the other hand, the impedance of underground secondary cables is generally low in comparison with the impedance of the transformers, making for a more equitable division of the load among all the transformers in the bank. In such cases, the total reserve capacity in the bank may be counted on rather than that in adjacent transformers only. The necessary reserve capacity, to a great extent, may be found in the overload capacity of the transformers for comparatively short periods of time, especially so on overhead systems. The sizes of fuses used with the transformers, however, must be chosen to allow for this emergency capacity.

Secondary banks are not very commonly used. They are more applicable to overhead systems, and are adapted to use in comparably light load density areas where more expensive reliable systems are not economically justified. Such a system does not provide for faults on the primary feeder supplying the bank.

SECONDARY NETWORKS

Where a very high degree of service reliability is desired, the secondary mains in that area are all connected together in a mesh or network, supplied from two or more "primary" sources (as compared to the secondary bank described above), Figure 2-3a. Because of their relatively much higher cost, they are usually confined to areas of high load density and underground systems. Such network systems may be either of the direct current or alternating current types.

Direct Current

The first electric systems were of the direct current type in the central downtown areas, starting in New York City, later introduced in Chicago, Cleveland, Detroit, Philadelphia, and other major cities. Originally of radial type supply, these systems developed into networks where the three wire 120/240 volt secondary mains were interconnected and the network supplied from low voltage feeders connected to the network at strategic points. The cable conductors are large and are not only expensive, but require large currents to burn clear secondary main cables in the event of fault. These supply cables emanate from a number of substations, relatively closely spaced in order to maintain reasonable levels of voltage in the network. The substations contain conversion equipment that changes high voltage alternating current supply feeders to low voltage direct current; these may be in the form of rotary converters or some type of rectifier. The chief advantage of the direct current system is that banks of storage batteries, usually located at substations, provide a reserve that insures power supply during relatively short periods of interruption of a substation (or even a generating station). Another lesser advantage stems from the better control of variable speed direct current motors (such as in elevators) as compared to alternating current motors. Additional loads, however, necessitate the installation of additional expensive conversion equipment, including at times additional substations.

The advent of induction and synchronous type alternating current motors has made practical the supplanting of direct current systems with alternating current systems. Because wholesale conversion of direct current systems is extremely expensive, such direct current systems have not been permitted to grow and have been gradually replaced with alternating current systems wherever practical. Direct current systems, although rapidly diminishing, exist in some areas and, hence, have been included in a minor way in this discussion.

Alternating Current

The low voltage secondary network is meant to provide, as far as practical, against service interruptions, even those of short or momentary duration. While the secondary bank generally provides for faults on transformers and secondary mains, it is vulnerable to faults on the single primary supply feeder, so that the reliability of service is no better than that of the primary supply feeder.

The secondary network provides against interruption of the primary supply feeders by supplying the network by more than one primary feeder. In the event of fault on one of the primary feeders, that feeder is automatically disconnected and the load is picked up by the other primary supply feeders. To prevent energizing the feeder in trouble from the secondary mains that remain energized from the other primary supply feeders, switches are placed between the secondary of each transformer and the secondary mains that make up the network. These switches, also known as network protectors, operate automatically when the current flow reverses its direction; should the network protector fail to open for any reason, a backup fuse, in series with the protector, is designed to blow, isolating that transformer from the energized secondary mains. The network protector will operate not only when trouble is experienced in the primary supply feeder, but also at times when voltage or phase angle differences between the supply feeder and the network is such that a reverse flow of energy will result.

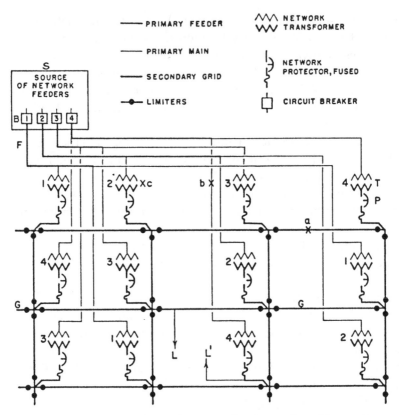

Figure 2-3a. One-Line Diagram of a Secondary Network
(From EEI Underground Reference Book)

Should the fault current and charging current on a feeder fail to blow all of the backup fuses at the network protectors, a short circuit and ground is applied to that feeder through a phantom or artificial load that controls the value of the short circuit current until the backup fuse or fuses blow, deenergizing that feeder.

Faults on secondary mains burn themselves clear. To limit the damage, fuses (known as limiters) are installed at the juncture of two or more secondary mains. Fault currents will operate these fuses, thus limiting the time the fault current persists, and limiting the damage from the burning at the fault in the secondary mains.

In the network supplied by a number of primary feeders, the transformers connected to these feeders are dispersed so that a transformer, deenergized when its feeder is deenergized, is surrounded by other

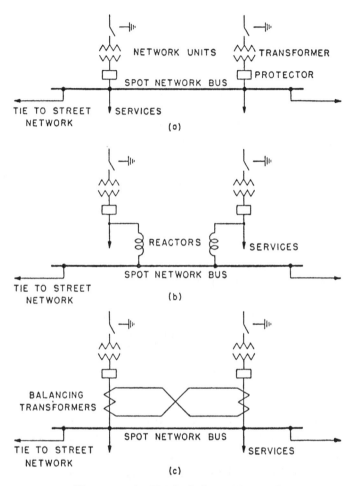

Figure 2-3b. Typical Spot Networks

(a) Two network units supplying a spot network bus from which services are tapped. (b) Two network units connected to spot-network bus through reactors. Services are supplied directly from terminals of network units. (c) Two network units supplying a spot-network bus through balancing transformers. Any of these forms of the spot network may or may not be connected to the secondary mains of a street network.

(Courtesy Westinghouse Electric Co.)

transformers that remain energized from their respective feeders. In this fashion, not only are overloads restricted on the supporting transformers, but adequate voltage is maintained on the secondary mains at and in the vicinity of the deenergized transformer.

The design of such low voltage networks limits the size of the network to values that permit it to be reenergized successfully by the simultaneous closing of the circuit breakers on all or a minimum number of feeders supplying the network. The procedure may call for blocking overcurrent relays temporarily during the closing of the circuit breakers; this provides for the momentary in-rush of current caused by the temporary loss of diversity among the loads on the feeders, as well as the charging currents associated with the cables of the feeders.

The capacity available at a service supplied by a number of transformers and the meshing of secondary mains in the vicinity enable both lighting and power loads to be served from the same service; this contrasts with separate lighting and power circuits often necessary with other types of secondary systems.

The low voltage secondary network is also employed in serving large individual loads requiring a high degree of reliability. These are known as spot networks, shown in Figure 2-3b. Note the use of secondary reactors in attempts to secure an equitable division of loads between the transformers supplied from the several primary feeders. Very tall office buildings requiring a similar degree of reliability also employ secondary network type of service; transformers are located in vaults on several floor levels and both primary and secondary cables are installed in vertical duct runs. In all cases, the secondary mains of these spot networks may be connected to the secondary mains in the street.

There remains to be determined the size of secondary mains. This must be considered in combination with the supply transformer. Essentially, the procedure includes the evaluation of several combinations in which annual charges on the cost of installation are compared to the annual cost of energy losses; that in which these two values are equal or nearly equal will be the most economical of standard sizes of transformers (in accord with Kelvin's Law). This equation may be written:

$$k_1 = k_2 + k_3$$

where k_1 = annual carrying charge of cost of transformer in place
         k_2 = annual cost of core loss
         k_3 = annual cost of conductor loss

For a given size of transformer, for practical purposes, k_1 and k_3 can be considered as constant regardless of the load. The conductor loss, however, will vary with the load:

$$k_3 = I^2 R$$

where I = load current in the secondary

$$= \frac{kW \times 1000}{E \times \text{power factor}} \quad (\text{for single phase})$$

kW = load in kilowatts

R = equivalent resistance (referred to secondary side) of transformer

t = equivalent hours (see above)

then $k_3 = \dfrac{I^2 R}{1000} \times t \times 365 \text{ days} \times C_3 \left(\dfrac{\text{cost of conductor}}{\text{loss}-/\text{kW hour}} \right)$

$$= \frac{kW^2 \times 1000^2 \times R \times t \times 365 \times C_3}{E^2 \times pf^2 \times 1000}$$

$$= \frac{365R \times t \times C_3}{E^2 \times pf^2} \times kW^2 = Q_1 \times kW^2$$

Total annual cost $= k_1 + k_2 + Q_1 \, kW$

Total annual cost per kW $= \dfrac{k_1 + k_2 + k_3}{kW}$

$$= \frac{k_1}{kW} + \frac{k_2 + k_3}{kW} = \frac{k_1 + k_2 + Q_1 kW^2}{kW}$$

_differentiating:

$$\frac{d(\text{total cost}/kW)}{d\,kW} = -\frac{k_1 + k_2 + Q_1 kW^2}{kW^2} = \frac{k_1 + k_2}{kW^2} + k_3 = 0$$

$$kW = \sqrt{\frac{k_1 + k_2}{k_3}} \text{ for least cost per kW}$$

The most economical load, therefore, is that for which $k_1 + k_2 = k_3$, that is, when the annual cost of conductor loss is equal to the annual charges of the installed transformer plus the annual cost of core loss. The standard size transformer equal to or nearest to this value will be the practical most economical size.

In determining the most economical size of the conductor associated with the transformer, the given load is assumed at a given distance from the transformer. The structures and facilities associated with the secondary conductors are also assumed to be the same for any size of the secondary conductor and will be neglected in the determination.

In the determination, the length of secondary is divided into two parts, one each way from the transformer, of length $\ell/2$ and load

$$\frac{D\ell}{2000}.$$

For uniformly distributed load, the loss is equal to what it would be if the load were all concentrated at one-third the distance from the transformer. Hence

$$I = \frac{\frac{D\ell}{2000} \times 1000}{E \times pf} = \frac{D\ell}{2E \times pf}$$

and

$$R_\ell = \frac{\frac{1}{3} \times \frac{\ell}{2}}{A}$$

where   I   = load current

D   = load density in kW per 1000 feet
ℓ = total length of secondary in feet
E   = circuit voltage
pf   = power factor of load
R_ℓ = resistance of circuit of length twice one-third of the total length $\ell/2$
r   = resistivity of conductor in ohms per mil foot
A   = cross sectional area of conductor in circular mils

then
$$I^2R = \frac{D^2\ell^2}{4E^2pf^2} \times \frac{r\ell}{3A}$$

Annual cost of losses

$$= 2\frac{I^2R}{1000} \times 365tC$$

$$= 2 \times 365 \frac{rD^2\ell^2tC}{3 \times 4 \times 1000AE^2pf^2}$$

and   t = equivalent hours (see above)

C = cost of conductor loss per kW hour

kc = annual carrying charges of three conductors in place, including the neutral

The factor 2 includes the total losses on the secondary on both sides of the transformer.

Let $Q_2 = \dfrac{2 \times 365rtC}{12 \times 10^3E^2pf^2}$ then the annual cost of losses

$$= Q_2\frac{D^2\ell^3}{A} \text{ and the total annual cost} = k_c\ell + Q_2\frac{D^2\ell^3}{A}$$

differentiating:

$$\frac{d(\text{total annual cost})}{dA} = k_c\ell - Q_2\frac{D^2\ell^3}{A} = 0$$

$$A = \sqrt{\frac{Q_2}{k_c}} \times D\ell$$

Like the size of transformers, the actual size of conductor will be the standard size equal to, or closest to, the annual cost of the conductor losses.

Other factors may (but probably will not) affect the final determination. There will be a practical minimum size of conductor because of the requirement of keeping voltage regulation (maximum and minimum values) within the prescribed limits as well as limiting in-rush currents to prevent voltage surges that may cause flicker. In the case of overhead lines, the conductor must be mechanically able to support itself. Secondary main conductors generally vary from (equivalent copper) sizes #6 to #4/0, which may be accommodated by the same structure and facilities; larger sizes may require reevaluation of the cost of conductors in place.

Secondaries without branches or loops have been assumed for simplicity. Where secondary banks and networks are concerned, the same methods may be employed by dividing such meshes into a series of radial pieces, as shown in the diagrams in Figure 2-4, without introducing appreciable error. Strict determination of current and voltage distribution in such meshes may be determined by application of several methods described in Appendix A.

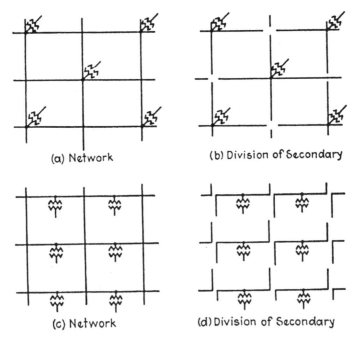

Figure 2-4. Division of Current in a Network
(Courtesy Westinghouse Electric Co.)

In the cases described above, single phase overhead circuits were assumed, but the same general methods may be applied to three phase (3 and 4 wire) circuits. For underground systems, because of the possibility of overheating of conductors, in determining conductor sizes, their capacitance should provide comfortable margins above the value of current to be carried. In the case of network design, consideration should also be given to the cable size for its ability to bum clear faults that may occur. These qualifying limitations may modify the determination of the economic size of the facilities constituting the secondary system.

This theoretical manner of determining transformer and secondary main conductor sizes is not often used in practice. The number of consumers connected to the secondary mains often is governed by the geographic arrangement of the area involved, or initially by limiting the number of consumers that would be affected by the deenergization of a transformer from whatever cause. The approximate total loads of these consumers is known and a transformer capable of serving this load is chosen. Placed in the center of the mains, the distance to the farthest consumer is known. The tolerable voltage drop to the last consumer is assumed and the resistance of the secondary main from that point back to the transformer is calculated. The conductor size matching this resistance (approximately) can be determined from tables of conductors and their characteristics.

The conductor sizes are indicated in this fashion for a number of secondary mains. From an inspection, a limited number of "standard" sizes that satisfy the greater number of these is chosen. Where the standard is greater than required to serve the consumers on some mains, the excess capacity not only accommodates load growth, but results in better voltage regulation and, in some instances, permits the installation of a smaller size transformer. Where the standards are too small to satisfy the requirements, the secondary main may be split into two or more pieces and separate transformers installed to serve each piece. The determination of distribution transformer standard sizes is accomplished in the same manner.

These procedures are economically justified from the resulting simplification of purchasing, stocking, installation, operation and maintenance practices, including the training and work practices of personnel and the selection and operation of equipment.

THE PRIMARY SYSTEM

The primary distribution system serves the load at higher than utilization voltages from sources (generally substations) to the point where the voltage is stepped down to the values at which it is utilized by the consumer.

The planning and design of primary circuits is considerably more difficult than for secondary systems as many more interrelated factors must be considered. For example, the determination of the most economical conductor size to carry a given load depends on the most economical load for a feeder, the division of loads between feeders and the most economical primary voltage. These are affected by the location and size of the supply substation, the arrangement of mains and laterals for polyphase circuits, including the selection of ungrounded, separately grounded or common grounding (with the secondary), all modified by the use of standard sizes of conductors, equipment, code and local restrictions and regulations. In the field of reliability, emergency connections with other circuits, loop circuits (open and closed), primary network, and throwover arrangements for individual consumers may be considered. Permissible voltage drop is affected by regulators (both at substations and on lines), capacitors, boosters. The choice of primary voltage is dependent not only on the availability of standard methods and equipment, but also on operating and maintenance procedures, and its impact on other utilities. And all are subject to further modification to provide for load growth and environmental requirements, and above all, for the safety of workers and the general public.

In determining the most economical bases of the several factors mentioned, for the sake of simplicity, certain assumptions are made: the loads distributed uniformly along the length of the primary line can be considered as a single load at its end, or to the feed point on the distribution circuit; the total energy loss in the line will be the same as if the total load were concentrated at one-third the distance to the end of the line; and the voltage drop over the line, to the end of the line, is the same as it would be if the total load were concentrated at the midpoint.

Economical Conductor Size

The choice of conductor size will be such that it will carry the load, that voltage regulation will be satisfactory, and that it will be mechanically strong to support itself on overhead lines or sustain installation stresses on underground systems. A standard operating voltage is assumed to determine the most economical size of conductor for that voltage.

From Kelvin's Law, the annual cost of energy loss in the conductor should equal the annual carrying charges on the installation. Hence, the annual carrying charges on poles and fixtures or ducts and manholes, as a percentage q_1 of cost $(k_1\ell)$ for conductor size up to about #2/0; for larger sizes, where heavier overhead construction or larger ducts and manholes may be required, part of this cost may be considered proportional to the conductor size $(k_2 + k_3)\ell$, where A = cross section of conductor in circular mils, k_1, k_2 and k_3 are unit construction costs, and ℓ is the length of the line or feeder in feet; the annual carrying charge q_2 on the conductor may be divided into two parts, one essentially constant for all sizes, and the other proportional to the cross sectional area; this applies to both manufacturing and installation costs:

$$\text{cost of wire} = (k_4 + k_5\,A)\ell$$
$$\text{cost of installation} = (k_6 + k_7\,A)\ell$$

where k_4 and k_5 are unit manufacturing costs and k_6 and k_7 are unit installation costs. Annual cost of operation, inspections, etc., may be considered a constant independent of conductor size, and is equal to $k_8\ell$. Energy costs are proportional to the square of the current carried and the resistance of the circuit:

$$I = \frac{kW \times 1000}{E \times pf} \text{ for single phase}$$

$$E = \frac{kW \times 1000}{3E \times pf} \text{ for three phase}$$

$$R = \frac{r\ell}{A}$$

where I = line current
       kW = load in kilowatts
        E = circuit voltage
       pf = power factor
        r = resistivity of conductor in ohms per mil-foot
        ℓ = length of conductor from source to load

Load at peak = I^2R watts per conductor

$$= \frac{I^2 r\ell}{A \times 1000} \text{ kilowatts per conductor}$$

$$= k_9 \frac{\ell}{A} \quad \text{where load, conductor material}$$
$$\text{and length are fixed}$$

where $k_9 = \dfrac{I^2R}{1000}$

and cost of energy loss $= \dfrac{k_9\ell}{A} \times 365tC$

where   C = cost of conductor loss per kilowatt-hour
        t = equivalent hours (see above)

The cost equations
$$q_1 (k_2 + k_3 A)\ell + q_2 (k_4 + k_5 A + k_6 + k_7 A)\ell + (k_8\ell)$$

$$= \frac{k_9\ell}{A} \times 365tC$$

where
$k_8\ell$ = cost of operation, inspection, etc.
k_{10} = $q_1 k_2 + q_2 (k_4 + k_6) + k_8$
k_{11} = $q_1 k_3 + q_2 (k_5 + k_7)$
k_{12} = $365tCk_9$

and $\left(k_{10} + k_{11}A\right)\ell = \dfrac{k_{12}\ell}{A}$

or total annual cost $(Y) = \left(k_{10} + k_{11}A + \dfrac{k_{12}}{A}\right)\ell$

differentiating, to obtain minimum cost,
$$\frac{dY}{dA} = \ell \, k_{11} - \frac{k_{12}}{A^2} = 0$$

$$A = \sqrt{\frac{k_{12}}{k_{11}}}$$

The selection of primary conductors and the primary voltage are obviously interrelated. The final selection is determined essentially by evaluating the carrying charges and energy losses on several primary systems operating at standard voltages. The most commonly used nominal primary voltages include:

System (volts)	Y System (volts)
2400	2400/4160
4800	4800/8300
7200	7200/12470
7620	7620/13200
13200	13200/23000
23000	20000/34500
34500	26000/45000

In evaluating the several primary voltage systems, factors other than conductor size must be included: substation facilities; line switches and insulators; in some cases, even pole lengths; distribution transformers and associated cutouts and surge arresters. At voltages higher than 13200 volts, operation and maintenance costs are affected as deenergization of conductors becomes necessary, or live-line methods must be employed in place of manual handling of conductors.

The most economical size of conductor for a given load can be determined by the study of annual costs and energy losses, as indicated above. Economic studies can also determine the theoretical optimum economical load for a given circuit, as well as the economical division of load among several feeders. Other factors, however, may operate to modify substantially the results of these economic studies.

The higher the primary voltage, the greater the load carrying ability of a particular circuit, while tolerable voltage drops are maintained. This may result in a large number of consumers supplied by a single feeder, and an undesirable number of consumers affected in the event of failure or deenergization of that feeder for whatever reason. The acceptable number of consumers that may be affected by an outage of the supply feeder is, in many cases, subject to nontechnical considerations depending on the importance of the consumers. Such limitation of the number of consumers will affect the size of conductor of the primary feeder main, generally smaller than that indicated by the original economic study. On the other hand, to improve the reliability of service in the service area, design features will provide for sectionalizing of the primary mains and switching arrangements that permit the reenergizing of unfaulted sections of a feeder from one or more adjacent feeders. The size of conductor chosen finally for the trunk or main primary line will therefore be large enough to carry the additional loads during periods of emergency, or when parts of a circuit are deenergized for construction or maintenance purposes.

For primary lines supplying one or more individual transformers, known as "laterals," the size of conductor is generally dictated by the mechanical properties of the conductor; for overhead systems it must be large enough to sustain itself in the spans between poles without undue sag or breakage; for underground systems, it must be large and strong enough to withstand the stresses on the cable being installed.

Voltage Regulation

The whole electric system is designed to deliver electrical energy to the consumer's service at a voltage that insures proper operation of all of the devices and equipment connected to that service. As the voltage will depend on the current flowing, and the value of current will vary with the consumer's demand from instant to instant, it is obvious that there will be a range, or voltage spread, that will satisfy the consumer's requirements. This voltage range is known as regulation and may be defined as the difference between the voltage at no load and that at full load, compared to that at full load; that is, the drop in voltage compared to that at full load. It is often expressed as a percentage. Most of the common lighting, motor, appliances, and other devices are designed to give satisfactory performance over a 10 percent range, usually 5 percent over and 5 percent under nominal rating; obviously, there are some types of service for which a closer regulation is desired and some that will tolerate a greater range.

This voltage regulation at the consumer's service will depend on the current flowing from the generator, through transmission lines, transformers and other substation equipment, primary circuits, distribution transformers, secondary mains and the service, Figure 2-5. Theoretically, this can be maintained by controlling the voltage at the generator terminals at the power plant. Occasionally, on a small independent system, this type regulation may be found; in practice, other means of attaining the objective are employed. Those associated with primary line voltages are described below.

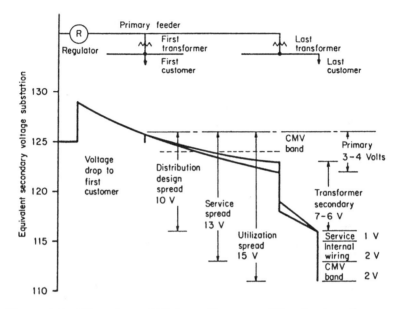

Figure 2-5. Allocation of Voltage Drops on Distribution Systems
(Courtesy Long Island Lighting Co.)

Transformer Taps

Satisfactory voltage regulation may be obtained by some fixed amount during both light and heavy load conditions. Taps on the distribution transformer can be changed, usually only on those on certain parts of the feeder. For example, on a feeder on which evenly distributed loads are assumed, the distribution transformers on the first third of the feeder from its source may be changed to lower the secondary voltage a fixed amount, those on the second third or middle part of the feeder may be left on their normal setting, while those on the third, farthest from the source, may have their taps changed to raise the secondary voltage a fixed amount. The effect of the tap change is to change the ratio of transformation from the normal to ratios above or below that normal, the changes usually reflecting a two- to five-percent change in the secondary voltage. The taps are usually made on the primary coil of the transformer, but may also be found on some secondary coils.

Boosters-Buckers

A similar effect may be obtained by the installation of a transformer that can boost or buck the line voltage a fixed amount. These are

usually employed on the laterals rather than on the trunk or main primary line. They may also be employed on the last portion of a primary
circuit, farthest from the source, usually in the boosting capacity.

Often, a standard distribution transformer is used for this purpose.
The primary and secondary coils are connected in series, essentially
operating as an autotransformer. The primary coil is connected across
the incoming primary circuit in the usual manner, while the outgoing
primary is connected between the terminal of the primary not connected
to the secondary and the terminal of the secondary coil. The voltage of
the secondary coil is either added to or subtracted from the incoming
primary voltage. (Figure 2-6)

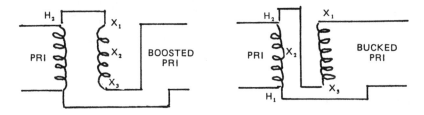

Figure 2-6. Boost-Buck Transformer Connection

Voltage Regulators

The voltage regulator acts very much as the booster-bucker autotransformer described above, except that the primary is wound
around a steel core that can rotate with reference to the stationary secondary coil. The voltage induced in the rotating primary depends on its
relative position to the secondary; the rotation in one direction causes
the voltage in the secondary to be added to the primary voltage, rotation
in the opposite direction subtracts the voltage.

Voltage regulators are most often located at the supply substation.
They may be used in two ways: individual regulators associated with
individual outgoing feeders; or one, larger in size, that regulates the
voltage on a bus from which two or more primary feeders are supplied.
Another application places a regulator (suitably constructed) on the far
end of a primary feeder to maintain satisfactory voltage regulation on
the feeder beyond this point.

A voltage regulator is rated according to the percentage it can add
to or subtract from the applied voltage. The capacity in kVA is based on
the product of full load amperes and the voltage that the regulator can

add or subtract. Regulators may be of single phase or of three phase operation; more will be discussed later under substations.

Capacitors

Voltage regulation may also be improved by the use of capacitors. Current flowing in the conductors of a circuit encounter not only resistance, but also the inductive reactance that is caused by the alternating magnetic fields of the conductors themselves and from those of adjacent conductors. It is measured in ohms. For sine wave alternating current, the inductive reactance may be expressed as:

$$\text{Inductive reactance} = X_L = \frac{2fL}{10^3} \quad \frac{\text{ohms per 1000 feet}}{\text{per conductor}}$$

where f = frequency in cycles per second

L = unit of inductance is the henry or millihenry (0.001 henry) and can be expressed by the equation

$$L = 0.1408 \left(\log_{10} \frac{b}{a} \right) + 0.0152 \, \mu$$

a = radius of conductor in inches

b = distance between centers of conductors in inches

p = permeability of materials (for copper or aluminum = 1; for steel = 13 to 16; copper or aluminum covered

steel

4 to 26 depending on size, stranding)

The voltage drop in a conductor caused by reactance = 1 × X_L where I is the current flowing in the conductor.

The effective voltage drop in the conductor is caused by both its resistance and reactance, but acting at right angles to each other. From the diagram, Figure 2-7, the resultant opposition to the flow of current, termed impedance (Z) can be found:

$$Z = \sqrt{R^2 + X_L{}^2}$$

If the inductance effect can be offset by capacitive reactance, the result

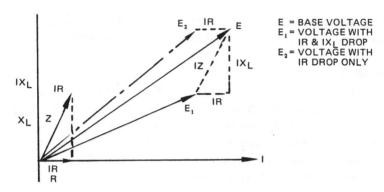

Figure 2-7. Effect of Reactance on Voltage Drop

of electron movement between adjacent conductors, the voltage drop can be limited (approximately) only to the I_R drop

$$X_C = \frac{10^6}{2\pi fC} \text{ ohms per 1000 feet of conductor}$$

where $C = \dfrac{0.007353}{\log_{10}\frac{b}{a}}$ microfarads per 1000 feet of conductor

This capacitance effect is usually too small to overcome the inductance effect, and for practical purposes, is usually neglected.

Voltage drop in an alternating current circuit depends on the resistance, reactance, current flow, and the power factor of the load on the feeder. What the capacitor does is to introduce capacitive reactance in the circuit to counter the effect of inductive reactance and have the power factor of the load on the feeder to equal or approach unity or 100 percent. The capacity of the capacitors (in kVA) and the location on the feeder where they are installed depends on how the loads are distributed on the feeder, the power factor of the loads, the feeder conductor size and spacing between conductors, and the voltage conditions along the feeder. Capacitors should be installed at or near the point where, during periods of heavy or maximum load, the voltage is at or below the minimum permissible voltage.

In an inductive circuit, the voltage drop may be expressed as:

$$v = I_R R + I_X X_L$$

connecting a capacitor in shunt on the individual circuit, the voltage drop then becomes:

$$v = I_R R + I_X X_L - I_C X_C$$

where v = voltage drop
    R = resistance of the line, in ohms
    X_L = inductive reactance of the line, in ohms (phase to neutral)
    I_R = power component of current in phase with voltage, in amperes
    I_X = reactive component of current lagging voltage by 90°, in amperes
    I_C = reactive component of current leading voltage by 90°, in amperes

X_L can be calculated from the equation above, or vectorially when the power flowing in a circuit is in kW and the power factor of the load is known, and the voltage.

Example: Assume a 7620 single phase primary voltage. The minimum voltage at a particular consumer or point in the feeder is 105 volts. The minimum voltage desired (as a standard) is 120 volts at heavy or peak load; the minimum correction required therefore is 15 volts. The maximum voltage at the point is 126 volts "at light load and the maximum voltage desired is 127 volts, Figure 2-8. Assume the reactance from the source to the capacitor point is 0.25 ohms.

$$I_C = \frac{E}{X_L} = \frac{15}{0.25} = 60 \text{ amperes (heavy load)}$$

and $\quad \frac{1}{0.25} = 4$ amperes (light load)

$= \quad 1 \text{ kVA} = \frac{1000}{7620} = 0.13 \text{ amperes/kVA}$

then Capacitor kVA $= \frac{60}{0.13} = 461$ kVA or (say) 10 – 50 kVA

units or 20-25 kVA units or 30-15 kVA units or combinations of these approximating 460 kVA.

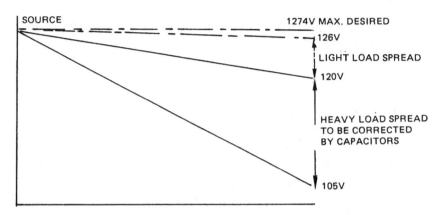

Figure 2-8. Voltage Improvement by Capacitors

At light load, the capacitors can only raise the voltage by 1 volt, then

$$\text{Capacitor kVA} = \frac{4}{0.13} = 31 \text{ kVA or (say) 2–15 kVA units}$$

that need to remain connected during light load periods. As the heavy load is reduced, capacitors may be disconnected to keep service voltage within permissible limits.

Shunt capacitors installed on a distribution feeder reduce the current, improve voltage regulation, and reduce energy losses in all parts of the system between the generators and capacitors, but have no effect beyond the point where they are installed.

Shunt Reactor

The discussion above is related to overhead systems. Where the primary circuit or feeder conductors are in a cable, the same effects take place, but because of the very close spacing of the conductors, the inductive reactance is relatively very small and is often neglected. If the cable has a metallic sheath, however, the conductors and sheath constitute a capacitor. When the cable length is comparatively long, as in the trunk or main of a primary circuit, and the primary voltage is fairly high (say 23 kV or higher), the capacitive effect is of a magnitude that the capacitive current (sometimes referred to as charging current) flowing in the conductor may be as large or larger than the current carrying capacity of the conductor. If the cable is to carry a load, not only will the conductors be overloaded, but the voltage of the circuit will be progressively larger as the distance from the source becomes greater. To remedy this condition, shunt reactors are connected to the conductor at strategic points along the feeder to counteract the capacitance effect and maintain circuit voltages at safe values and within permissible limits. The kVA size of reactors is calculated in the same manner as that of capacitors. Only few instances of this effect on primary cable distribution circuits exist.

Series Capacitors

Capacitors installed in series in a primary distribution circuit also act to regulate the voltage of the circuit. The current flowing through them continually produces a capacitive effect that acts to counteract the inductive effect of the load current flowing in the feeder. Moreover, the effects vary with the load current, and hence compensation varies automatically as the current changes, eliminating the necessity to switch capacitors on and off as is the case with shunt capacitors. Further, the voltage on the load side of the series capacitor raises the voltage above the source side, and does so instantaneously, acting faster than regulators or switched shunt capacitors and without the associated control devices.

A serious disadvantage, however, arises when relatively large fault currents flow through the capacitor, resulting in voltage rises that may destroy the capacitor and cause damage to adjacent facilities. It is necessary, therefore, to install automatically operated by-passes to short circuit the capacitor as quickly as possible. The potential hazard limits this type of relatively rare installation usually to large single consumers where flicker or rapid and repeated voltage fluctuations caused by frequent motor starts, welders, furnaces, and similar loads may be objectionable and intolerable (e.g., effect on computers, strobe lights, etc.).

Capacitors and reactors may also be applied to secondary systems but their employment in this fashion is extremely rare.

Resonance

When the capacitance effect in a circuit exactly balances the inductive effect so that the net effect is zero, the only quantity left is the resistance. When this occurs, the circuit is said to be in resonance.

Polyphase Primary Systems

In the discussion so far, single-phase, two- and three-wire, circuits have been used for simplification in describing the functioning of supply circuits. Single-phase circuits are adequate in supplying relatively small loads, including lighting and small or fractional horsepower motors. For larger bulk loads and the supply to larger motors, polyphase systems are more economical and provide smoother operation of motors. The polyphase systems may be two-phase systems (that are becoming rapidly obsolete, although some will continue to exist) and three-phase systems that are almost universally employed in transmission and distribution systems. For purposes of comparison, the single-phase system will be used for reference. Comparison of the characteristics of such systems is shown in Appendix D.

Single-Phase System

The two-wire system, Figure 2-9, consists of a two-conductor circuit with constant voltage maintained between the conductors, the load being connected in parallel across the circuit. For low-voltage lines, one conductor is usually grounded. The grounding of one conductor, usually called the neutral, is a safety measure. Should the live conductor accidentally come in contact with the neutral conductor, the electrical pressure or voltage of the live conductor will be dissipated over a relatively large body of the earth and thereby rendered harmless. In figuring power loss (I^2R) and voltage drop (IZ), the resistance and reactance, or impedance, of both conductors must be considered.

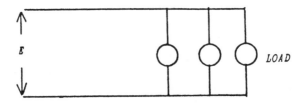

Figure 2-9. Alternating Current, Single-Phase, Two-Wire System
(Courtesy Long Island Lighting Co.)

The three-wire system, Figure 2-10, is the equivalent of two two-wire systems combined so that a single wire serves as one wire of each of the two systems, the neutral. At a given instant, if one outer wire is E volts (e.g., 120 volts) above the neutral, the other will be E volts (also 120 volts) below the neutral, and the voltage between outside wires will be 2E (or 240 volts).

If the load is balanced between the two systems, the neutral conductor will carry no current and the system acts as a two-wire system at twice the voltage of the component system with each unit of load (such as a lamp) of one component system in series with a similar unit of the other system. If the load is not balanced, the neutral conductor will carry a current equal to the difference between the currents in the outside conductors. For low voltage lines, the neutral is usually grounded. For a balanced system, power, power loss and voltage drop, are determined in the same way as for a two-wire circuit consisting of the outside wires; the neutral is neglected.

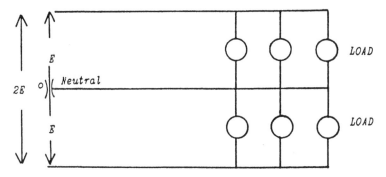

Figure 2-10. A-C Single-Phase, Three-Wire System
(Courtesy Long Island Lighting Co.)

An unbalanced condition in the three-wire circuit is indicated in Figure 2-11. Let the distance between the broken line represent the voltage. There will be a drop in voltage toward the neutral in both conductors I and 2. The neutral conductor now carries the difference in currents, that is, $1_2 - 1_1 = I_n$; this current in the neutral will produce a voltage drop in the neutral, as shown in the figure. The result will be a much larger drop in voltage between conductor 2 and neutral than between conductor 1 and neutral. If the unbalance is large, I_n being greater than I_1, E_{R_1} will be greater than E_{S_1} or there will be a rise in voltage across that side.

The limiting case zoccurs when $I_1 = 0$ and $I_n = I_2$. In that case all the load is carried on side 2; the rise in voltage on side I will be half as much as the drop in voltage on the loaded side. If an equal load is now added on side 1, the load being balanced, $I_n = 0$ and the drop in voltage between 2 and neutral is only half that obtained with load on side 2 only, although the total load is now doubled. In all of the discussion, it has been assumed that the size of the neutral conductor is the same as the outside conductors.

Two-Phase Polyphase System

The four-wire system, shown in Figure 2-12, consists of two single-phase, two-wire systems in which the voltage in one system is 90° out of phase with the voltage on the other system, both being supplied from the same source. In determining values of power, power loss, and voltage drop in such a system, the values are calculated as two separate single-phase, two-wire systems.

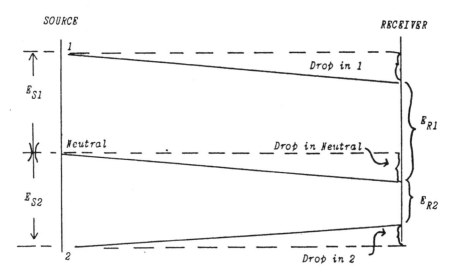

Figure 2-11. Unbalanced Load, Single-Phase, Three-Wire System
(Courtesy Long Island Lighting Co.)

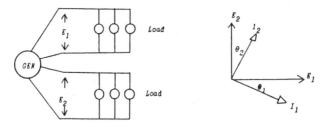

Figure 2-12. A-C, Two-Phase, Four-Wire System—and Vector Diagram
(Courtesy Long Island Lighting Co.)

The three-wire system, shown in Figure 2-13, is equivalent to a four-wire two-phase system with one wire made common on both phases. The current in the outside or phase wires is the same as in the four-wire system; the current in the common wire is the vector sum of these currents but opposite in phase. When the load is exactly balanced on the two phases, these two currents are equal and 90° out of phase with each other, and the resultant neutral current is equal to $\sqrt{2}$ or 1.41 times the phase current. The voltage between phase wire and common wire is the normal phase voltage, the same as in the four-wire system. The voltage between phase wires is equal to $\sqrt{2}$ or 1.41 times that voltage. The power transmitted is equal to the sum of the power transmitted by each phase. The power loss is equal to the sum of the power losses in each of the three wires.

The voltage drop is affected by the distortion of the phase relation caused by the larger current in the third or common wire. In Figure 2-13, if E_1 and E_2 are the phase voltages at the source and I_1 and I_2 the corresponding phase currents (assumed balanced loading), I_3 is the current in the common wire. The impedance drop over each conductor (IZ) subtracted from the voltages shown give the resultant voltages at the receiver of AB for phase 1 and AC for phase 2. It is apparent that these voltage drops are unequal and that the action of the current in the common wire is to distort the relations between voltages and currents, although the effect shown in the figure is exaggerated.

The five-wire system, shown in Figure 2-14, is equivalent to a two-phase four-wire system with the middle point of each phase brought out

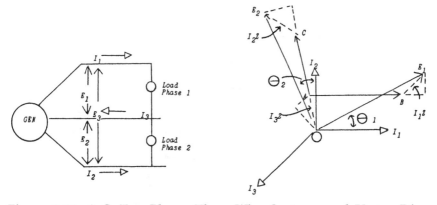

Figure 2-13. A-C, Two-Phase, Three-Wire System—and Vector Diagram. (*Courtesy Long Island Lighting Co.*)

and joined in a fifth wire or common wire. This value may be in the nature of 120 volts and may be used for lighting and small power loads, while the voltage between the pairs of phase wires, E, may be 240 volts and used for larger loads. If the load is exactly balanced on all four phase wires, the common wire or neutral carries no current. If not balanced, the neutral carries the vector sum of the unbalanced currents in the two phases.

Three-Phase Polyphase System

The four-wire system, Figure 2-15, is equivalent to three single-phase systems supplied from the same generator, the voltage in each phase being 120° out of phase with the other two phases. One conductor is used as a common conductor for all three systems. The current in that common or neutral conductor is equal to the vector sum of the currents in the three phases but in opposite phase, I_n in the figure. If these currents are nearly equal, the neutral current will be small since these phase currents are 120° out of phase with each other. Usually, the neutral is grounded. Single-phase load is usually connected between one phase and the neutral, but may be connected between phase wires if desired. In this latter case, the voltage is $\sqrt{3}$ or 1.732 times the line to neutral (E) voltage. The separate phases of a three-phase load may similarly be connected either way. Power transmitted is equal to the sum of the power in each of the three phases. Power loss is equal to the sum of the losses in all four wires.

When there is a neutral current, the voltage drop is affected by the distortion of the phase relations due to the voltage drop in the neutral conductor. When the neutral conductor is grounded at both the sending

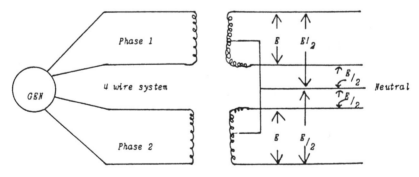

Figure 2-14. A-C, Two-Phase, Five-Wire System
(Courtesy Long Island Lighting Co.)

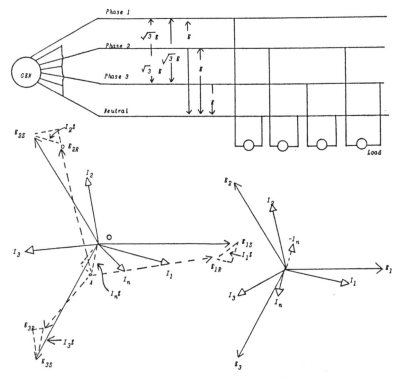

Figure 2-15. Voltage Drop in Three-Phase, Four-Wire System
(Courtesy Long island Lighting Co.)

and receiving ends, the neutral drop is theoretically zero, the current returning through ground. The voltage drop may be obtained graphically, as shown in Figure 2-15, by applying the impedance drop of each phase to its voltage. The neutral point is shifted from 0 to A by the voltage drop in the neutral conductor and the resulting voltages at the receiver are as shown by E_{1R}, E_{2R} and E_{3R}. The voltage drops in each phase are numerically equal to the difference in length between E_{1S} and E_{1R}, E_{2S} and E_{2R}, and E_{3S} and E_{3R}.

In the three-wire system, Figure 2-16, if the load is equally balanced on the three phases of a four-wire system, the neutral carries no current and could be removed, making a three-wire system. It is not necessary, however, that the load be exactly balanced on a three-wire system. Considering balanced loads on a three-wire system, three-phase loads may be connected with each phase connected between two phase wires in a delta

(Δ) connection, or with each phase between one phase wire and a common point in a star or Y connection, as shown in the figure.

The voltage between line wires is the delta voltage, E_Δ, while the line current is the Y currently. The relations in magnitude and phase between the various delta and Y voltages and currents for the same load is shown in the figure. For delta connection, l_Y is equal to the vector difference between the adjacent delta currents, hence:

$$I_Y = \sqrt{3} \text{ or } 1.732I\Delta \qquad \text{and } E_\Delta = \sqrt{3} \text{ or } 1.732E_Y$$

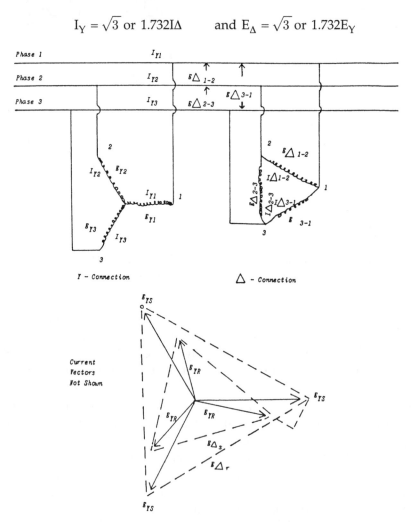

Figure 2-16. A-C, Three-Phase, Three-Wire System
(Courtesy Long Island Lighting Co.)

Power transmitted, when balanced loads are considered, is equal to three times the power transmitted by any one phase. Power loss is equal to the sum of the losses in each phase, or when balanced conditions exist, to three times the power loss in any one phase.

Voltage drop in each phase, referred to Y voltages, may be determined by adding the impedance drop in one conductor vectorially to E_Y, when balanced loads are considered. The same thing is done in determining voltages when unbalanced loads are considered. If E_S is the voltage between phases at the source, E_{YS} the phase to neutral voltage, the IZ drop is subtracted vectorially from E_{YS} for each of the three phases and the resulting voltages between phases at the receiving end $(E_{\Delta R})$ obtained.

Transformer Connections

Before discussing the several types of connection of distribution transformers, it is desirable that the subject of transformer polarity be understood.

Polarity

The relative direction in which primary and secondary windings of a transformer are wound around the core determines the relative direction of the voltages across the windings. The "direction" of the winding depends on which end is used as the starting point. For a single-phase transformer, if the direction of the applied voltage at any instant is assumed as from a to b in Figure 2-17, the direction of the voltage across the secondary circuit will be either from c to d, or from d to c, depending on the relative direction of the windings.

Since it is essential, if two transformers are to be paralleled, to know the relative direction of the voltages of the two transformers, certain conventions have been established for designating the polarity of a transformer. The designation of polarity may be illustrated by the figure, if one high-voltage load is connected to the adjacent opposite low-voltage terminal, (say) a to c, the voltage across the two remaining terminals, b and d, is either the sum or difference of the primary and secondary voltages, depending on the relative directions of the windings. If the voltage b to d is the sum, the transformer is said to have additive polarity; if it is the difference, the transformer is said to have negative polarity. If the vectors E_{ab} and E_{cd} are in phase, the polarity is subtractive; if they are in phase opposition, the polarity is additive. To indicate

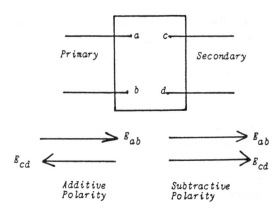

Figure 2-17. Polarity of a Transformer
(Courtesy Long Island Lighting Co.)

whether the transformer is of additive or subtractive polarity, the terminals are marked as shown in Figure 2-18.

To connect two secondaries in parallel, similarly marked terminals
are connected together irrespective of the polarity, provided similarly
marked primary terminals have been connected together. In general,
standard distribution transformers usually are of additive polarity, while
substation transformers are of subtractive polarity.

Single-Phase Transformer Connections

A typical arrangement of the method of bringing the terminals of
the primary and secondary coils out through the tank of a single-phase
distribution transformer is shown in Figure 2-19. To provide for flexibility of connection, the primary and secondary coils are usually each ar-

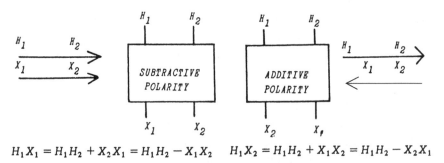

$$H_1X_1 = H_1H_2 + X_2X_1 = H_1H_2 - X_1X_2 \qquad H_1X_2 = H_1H_2 + X_1X_2 = H_1H_2 - X_2X_1$$

Figure 2-18. Polarity Markings of Single-Phase Transformer
(Courtesy Long Island Lighting Co.)

ranged in two sections, each section of a coil having the same number of turns and, consequently, generating the same voltage. The two primary sections are usually connected together inside the tank and only two primary terminals are brought out from one side of the tank through bushings which insulate them from the tank casing. Four secondary leads are similarly brought out through insulating bushings from the opposite side of the tank, two terminals are brought out from each half of the secondary coil. The two inner connections from this secondary are transposed before being brought out through the casing. A schematic diagram of this arrangement is shown in the figure.

Three different methods of connecting such a transformer for operation are shown in Figures 2-19a, b, c. In Figure 2-19a, the two sections of the secondary coil are connected in parallel to supply a two-wire 120-volt secondary circuit. In Figure 2-19b, the two sections are connected in series to supply a 240-volt circuit. The connection in Figure 2-19c is that to supply a three-wire 120/240-volt circuit; the third wire is connected to the midpoint of the secondary coils, permitting 120- and 240-volt apparatus to be supplied.

Two-Phase Transformer Connections

Two-phase transformation of power is usually made with two single-phase transformers connected as shown in Figures 2-20a, b, and c for three-, four-, and five-wire secondary circuits, respectively.

Three-Phase Transformer Connections

Three-phase transformations may be accomplished by means of three single-phase transformers or by a three-phase transformer. The methods of connecting the coils for three-phase transformation are the same whether the three coils of one three-phase transformer or the three coils of three separate single-phase transformers are used. The most commonly used connections are the delta and the star or Y connection, explained earlier. With these two types of connections, four combinations are available:

1. Primary connected in Y   Secondary connected in Y Figure 2-21

2.   "   "   " Δ   "   "   " Δ Figure 2-22

3.   "   "   " Y   "   "   " Δ Figure 2-23

4.   "   "   " Δ   "   "   " Y Figure 2-24

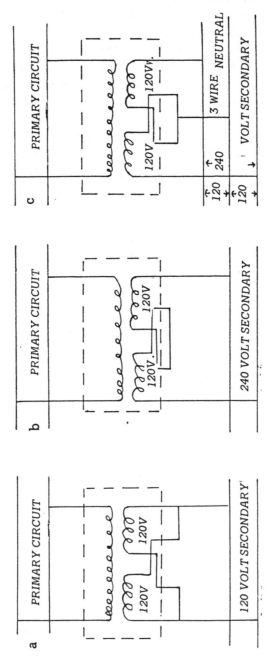

Fig. 2-19a. Single-Phase Transformer Connected for Two-Wire, 120-Volt Secondary.
Fig. 2-19b. Single-Phase Transformer Connected for Two-Wire, 240-Volt Secondary.
Fig. 2-19c. Single-Phase Transformer Connected for Three-Wire, 120/240-Volt Secondary.

Figure 2-19. Single-Phase Transformer Connections

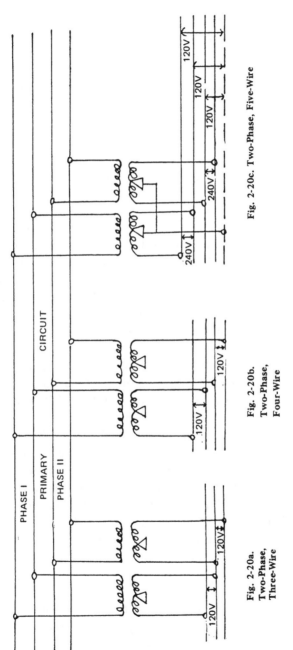

Fig. 2-20a.
Two-Phase,
Three-Wire

Fig. 2-20b.
Two-Phase,
Four-Wire

Fig. 2-20c. Two-Phase, Five-Wire

Figure 2-20. Connections for Two-Phase Transformers

Y- Y Connection

With the Y-Y connection, the secondary circuit is in phase with the primary circuit and the ratio of primary to secondary voltage is the same as the ratio of turns in the phases. The secondary may be either three-wire or four-wire as desired. Unbalanced currents on the secondary circuit are transmitted through the transformers to the primary unchanged in phase relation, although reduced in magnitude according to the ratio of the windings.

Delta-Delta Connection

The delta-delta connection does not cause any phase angle shift between primary and secondary, nor does the ratio of transformation of the bank differ from the turn ratio of the windings. The phase voltage is 1.73 times the line to neutral voltage of the Y-Y bank and each phase requires more turns on the winding than the Y-Y bank. The phase current under balanced load is only 0.57 times the line current. Unbalanced loads tend to divide more evenly among the phases than with a Y-Y bank. With a single-phase load, for example, the current divides among the windings according to the relative impedance of each path, two-thirds on the nearest phase and one-third on the others, if the phases have the same impedance.

Figure 2-25a shows the effect of a single-phase load on a Y-Y bank and Figure 2-25b shows the same load on a delta-delta bank. For convenience only one set of windings is shown and the delta is opened up to show the currents more clearly. In Figure 2-25a, it will be noted that the current from a single-phase line-to-line load has its phase angle lag reduced by 30° in phase A and increased by 30° in phase B. In the delta in Figure 2-25b, phase A has two-thirds of the current at normal power factor, while in phase B the angle of lag of the current is increased by 60°, and in phase C it is reduced by 60° and is actually leading. With this connection, there is also an advantage in case of damage to one of the transformers, the bank can be operated with only two of the transformers as an emergency condition.

Y-Delta Connection

In the Y-delta connection, there is a 30° phase angle shift between primary and secondary. This phase angle difference can be made either leading or lagging, depending on the external connections of the transformer bank. The relation between the high voltage and low voltage

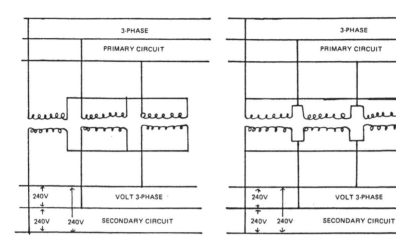

**Figure 2-21. Three-Phase
Transformation;
Wye-Wye Connection**

**Figure 2-22. Three-Phase
Transformation;
Delta-Delta Connection**

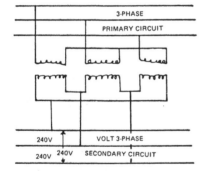

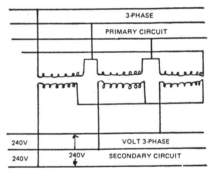

**Figure 2-23. Three-Phase
Transformation;
Y-Delta Connection**

**Figure 2-24. Three-Phase
Transformation;
Delta-Y Connection**

(All figures courtesy of Long Island Lighting Co.)

sides is shown in Figure 2-26. The phase rotation may be either way, the diagram merely shows the phase position of the windings with respect to each other. In a Y-delta vector diagram, the phase vectors of primary and secondary must be drawn parallel to each other.

With balanced load the power factor of the primary circuit is the same as the power factor of the secondary circuit. The transformation

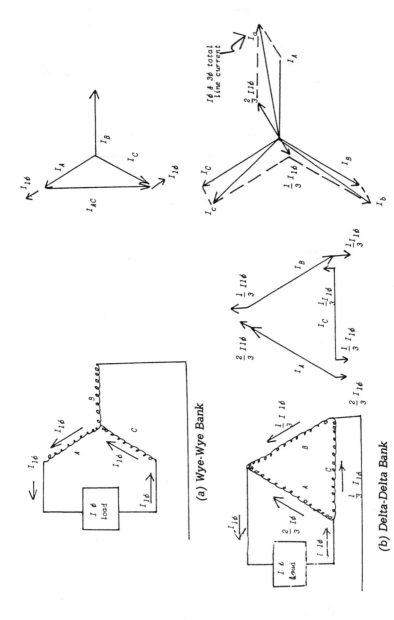

(a) Wye-Wye Bank

(b) Delta-Delta Bank

Figure 2-25. Vector Diagram for Single-Phase Load on Wye-Wye Bank or Delta-Delta Bank. *(Courtesy Long Island Lighting Co.)*

ratio, however, is not the same as the ratio of turns of the phases. For example, three 2400/240-volt transformers connected in Y on the primary side and delta on the secondary side give a transformation ratio of 4160/240 volt or 1.73 to 1, although the transformers are themselves 10 to I in ratio. At the same time, the secondary line current is 17.3 times the primary current. Under unbalanced loads, the current and power factor in each phase of the primary is different from the corresponding relation in the secondary circuit. The determining factor is that with the primary neutral floating, that is, there is no outlet for the neutral current, the primary phase currents must add up to zero.

The single-phase load in Figure 2-26 divides so that B phase carries two-thirds of the current, while A and C phases carry one-third. The volt-ampere capacity required in the transformer and the primary feeder is therefore four-thirds the actual load in volt-amperes. If a second single-phase load is connected to another leg of the delta, the corresponding load in the primary phases may be independently determined, and the resultant primary currents vectorially combined with the preceding results to give the answer for both loads.

Delta- Y Connection

The same discussion applying to the Y-delta connection applies also to this connection, the positions of primary and secondary being reversed.

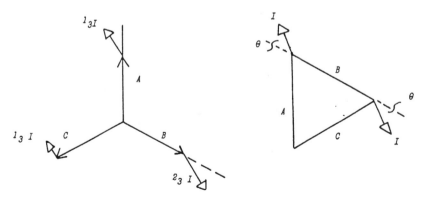

Figure 2-26. Vector Diagram for Single-Phase Load on Wye-Delta Bank. (*Courtesy Long Island Lighting Co.*)

Polarity of Three-Phase Transformers

Figure 2-27 shows the order of bringing out and marking transformer terminals in a three-phase transformer. In the determination of the polarity of three-phase transformers, consideration of phase rotation, marking of terminals, and types of internal connection are involved. Considering first the delta-delta or Y-Y connections: For these two connections, the corresponding line voltages may have either a 0° or 180° displacement. With 0° displacement, if the two corresponding terminals are connected together (say) H_1 and X_1, then H_1 H_2 is in phase with X_1 X_2, and H_2 H_3 with X_2 X_3, etc. With 180° displacement, on the other hand, H_1 H_2 will be in phase opposition to X_1 X_2, etc. The vector diagrams for 0° and 180° displacement are shown in Figure 2-28.

In the case of Y-delta connection, when two corresponding terminals of the high and low side are connected together, the other corresponding voltages H_1 H_2 and X_1 X_2, H_2 H_3 and X_2 X_3, H_3 H_1 and X_3 X_1, may be 30° leading or lagging from each other. The vector diagrams for Y-delta or delta-Y connections, for this condition, are shown in Figure 2-29a. In Figure 2-29b is shown the vector diagram for a 30° lag of Y side to delta side. It will be noted in Figure 2-29a, if X_1 is placed on H_1 the voltages X_1 X_2, X_2 X_3, and X_3 X_1 will lead H_1 H_2, H_2 H_3, and H_3 H_1 by 30° for a delta-Y connection and lag 30° for the Y-delta connection. The reverse is true for Figure 2-29b.

Making Delta Connection

In completing the delta connection, precautions should be taken that the third transformer is connected so that its voltage will have the proper phase relation to the other two, otherwise a dangerously high

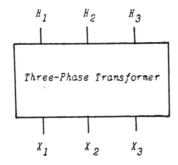

Figure 2-27. Arrangement of Loads on a Three-Phase Transformer
(Courtesy Long Island Lighting Co.)

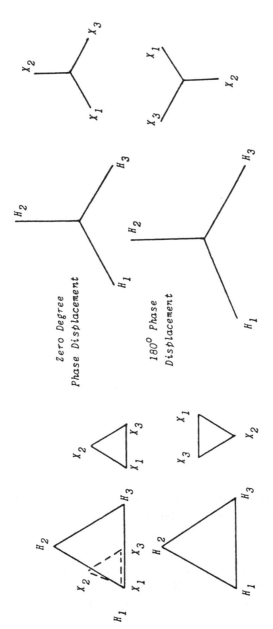

Figure 2-28. Polarity of Three-Phase Transformer Showing Phase Displacement. (Courtesy Long Island Lighting Co.)

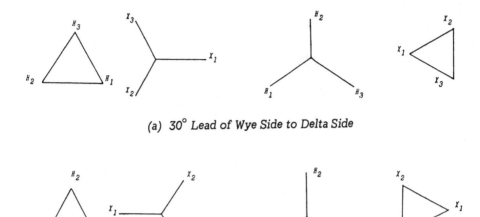

(a) 30° Lead of Wye Side to Delta Side

(b) 30° Lag of Wye Side to Delta Side

Figure 2-29. Arrangement of Terminals in Y-Delta Connection
(Courtesy Long Island Lighting Co.)

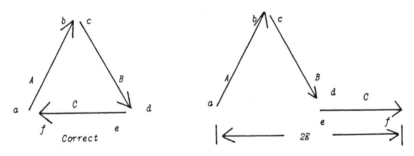

**Figure 2-30. Correct and Incorrect Method of Making Delta Connec-
tion** *(Courtesy of Long Island Lighting Co.)*

voltage may be obtained across this phase. This condition is indicated in
Figure 2-30. Before points a and f are connected together, a test should
be made by voltmeter or equivalent apparatus to be sure that the voltage
between points a and f is very close to zero. Potential transformers may
be necessary for this test.

Other Three-Phase Connections

Open Delta Connection. The open delta connection is the result of omitting one transformer from a closed delta connection. Referring to Figure 2-31, if one of the transformers is removed, the other two would still maintain the correct voltage and phase relations on the secondary. But, in a closed delta connection, each transformer carries $1/\sqrt{3}\, I_2$ current at E_L volts. If one transformer is disconnected, the current in each of the remaining two transformers would be I_2 amperes instead of $1/\sqrt{3}$ I_L. Since the line current must not exceed the rated current of a transformer if it is not to be overloaded, the load must be reduced. Thus, there is not only a reduction to 66-2/3 percent of the original closed delta transformer capacity, but also a reduction of

$$\frac{1}{\sqrt{3}}$$

or 57.7 percent of the line current to prevent damage to the transformer windings. Since the line current and the coil current in an open delta connection are the same, then:

$$\frac{57.7\% \text{ current}}{66.7\% \text{ transformer capacity}} = 86.6\% \text{ rating}$$

which means that when two transformers are connected in open delta, they must operate at 86.6 percent of their rating if they are not to be overloaded.

Open Y-Open Delta Connection. To avoid the case of three transformers for small three-phase loads, this connection is used. The diagram of connections and associated vector diagrams are given in Figure 2-32.

Voltage and Phase Transformations

Three-Phase Primary to Two-Phase Secondary

The transformation from three phase to two phase is accomplished by using one Y-type and one delta-type transformer, as shown in Figure 2-33; that includes 4160 volt and 2400 volt transformers as an example. The 4160 volt transformer connected between lines A and B, and the 2400 volt transformer connected between line C and neutral. These con-

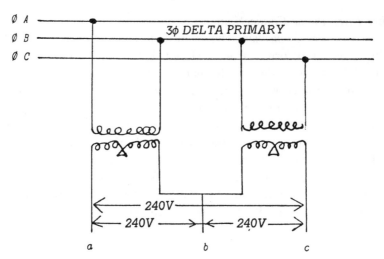

Figure 2-31. Open Delta Connection
(Courtesy of Long Island Lighting Co.)

nections give a 900 displacement between the two phases, as shown in the vector diagram.

Scott or T Connection

A distribution transformer having a standard 86.6 percent tap may be connected with one having a 50 percent tap as shown in the diagram of Figure 2-34 to obtain a two-phase supply from a three-phase source, or vice versa. The associated vector diagram is also included in the figure.

Autotransformer Connection

An autotransformer may also be used to provide a two-phase supply from a three-phase source; the connections and vector diagram are shown in Figure 2-35.

Primary Circuit Reliability

The radial-type primary circuit is the most frequently employed in electric distribution systems. A radial circuit, so named because it radiates from its substation source and traverses the area to be supplied. Transformers may be connected to the feeder main along the trunk route and laterals, both single- and three-phase, may be tapped off of the feeder main to supply loads not adjacent to the main. More often, the

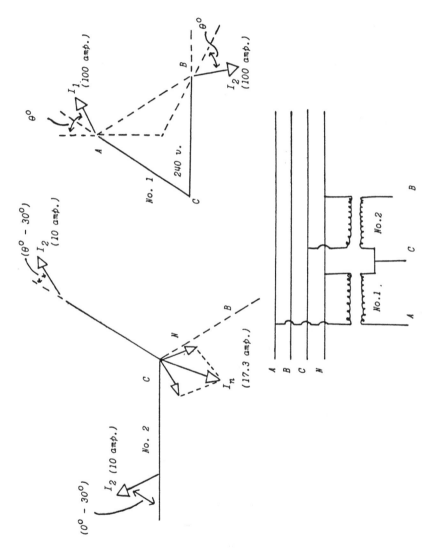

Figure 2-32. Open-Wye Open-Delta Connection
(Courtesy Long Island Lighting Co.)

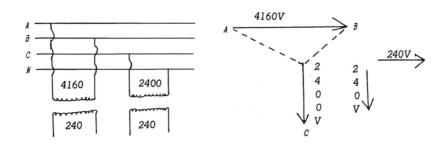

Figure 2-33. Three-Phase Four-Wire Primary to Two-Phase Four-Wire Secondary. *(Courtesy Long Island Lighting Co.)*

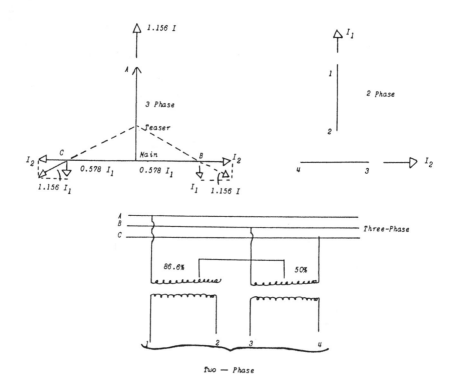

Figure 2-34. Scott or "T" Connection
(Courtesy Long Island Lighting Co.)

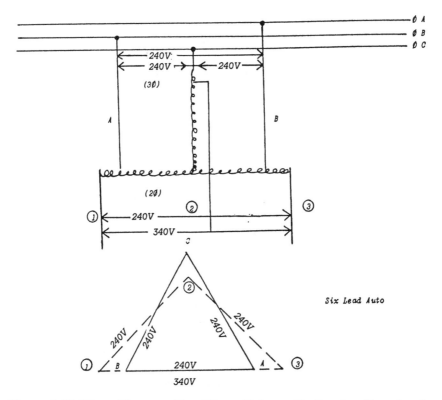

Figure 2-35. Three-Phase or Two-Phase Using a Six Lead or Four-Lead Autotransformer. (*Courtesy Long Island Lighting* CO-)

total area to be supplied may be divided into three separate phase areas in which all the single-phase loads are supplied from the same phase. The areas are so selected that the loads on each phase are about the same in order that loads (and voltages) be balanced on the three-phase circuit.

As modification of this feeder design, a "super" main runs untapped directly from the source to the load center of the area served, from which point (three phase) mains radiate in all directions. The load capacity of such a circuit may be several times greater than the simple radial circuit described above. The size of the conductor of the mains radiating from the load center may be smaller because of the much lower load density.

Both circuit arrangements are designed so that, in the event of fault, the circuit can be sectionalized allowing the unfaulted parts of the circuit to be energized from alternate sources, usually adjacent circuits.

To permit this type of operation, the initial loading of each circuit (at peak) should be limited to approximately two-thirds of the potential capacity of each circuit.

Consumers, usually large, requiring a better reliability, are also provided with an alternate source. Primary lines from two separate feeders are brought to a throw-over device that permits selection of service from either source; the throw-over arrangement may be operated manually or automatically.

Another design, usually employed on long and widely extended circuits, calls for reclosers to be installed on one or more branches of a circuit so that a fault on the branches need not affect the entire circuit. Temporary faults, such as a tree limb falling on a line, will not cause a lengthy outage to service on the branch, but only a momentary interruption. (Figure 2-36)

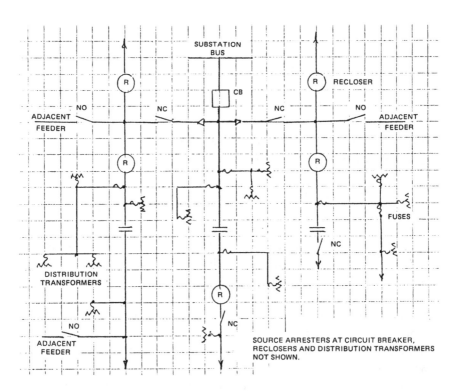

Figure 2-36. Radial Primary Feeder with Protective and Sectionalized Devices.

Loop Primary Circuits

A high degree of service continuity may be realized making use of the alternate feed principle with so-called loop systems. Here, the circuit, as its name implies, forms a loop, starting at one source and returning to the same or other source, with sectionalizing devices installed on both sides of a single transformer location or on both sides of a group of transformers; approximately midway, or at one end, of the loop, another sectionalizing device, usually a circuit breaker, is installed. The sectionalizing devices may be manually operated switches, or more expensive automatically operated circuit breakers, or a combination of the two. Such systems essentially provide a two-way primary feed to the high reliability consumer.

A fault on any part of this circuit will deenergize the entire circuit until the sectionalizing switches between the fault and the station breaker can be opened (possibly by automation); the circuit breaker at the station will reclose (if programmed to do so) and service restored to those consumers from the point of fault back to the station on the remaining part of the loop. The process is repeated after the fault is cleared; the entire circuit is again deenergized by the station breaker until the open sectionalizing switches are closed, then the station breaker is closed, restoring service to the entire circuit.

In Figure 2-37b, the switches are replaced with the much more expensive circuit breakers and associated relaying. Since all the switching necessary to clear the fault is done automatically, any interruption may only be momentary.

While this closed loop arrangement appears to be very reliable, there is danger that the fault may be so located that the fault current flowing in one branch of the circuit from the fault may not be great enough to operate the isolating breaker in that branch of the loop. By operating the loop in two sections, that is, as an open loop, total fault current will flow to the fault located between the open (loop) circuit breaker and the station, assuring the operation of the circuit breaker nearest the fault between the fault and the station. Meanwhile, the open circuit breaker (0 in the diagram) sensing one side is now deenergized, will close automatically, restoring the open loop and service to all the consumers on the circuit, except those connected to the isolated section on which the fault is located.

In Figure 2-37c, the two generators may represent two separate generating stations distanced apart, then the loop represents a tie between the two generators One of the breakers may operate normally open. Should it be found desirable to operate with it closed, precautions should be taken to synchronize the two generators before closing the circuit breaker.

Should a fault develop on this now tie circuit, fault current would flow from each of the generators in proportion to the distance each is from the fault. This could result in slowing down the unit supplying most of the fault and load current, and a relative speeding up of the unit furthest from the fault. The result would be a continuous rocking motion between the two generators, accelerating, so that ultimately one would have its circuit breaker open from overload, and subsequently the other would also have its circuit breaker open, resulting in this part of the system to shut down. The same effect would take place with a number of generators similarly distanced and connected together. The result would be that each generator would have its circuit breaker open, "cas-

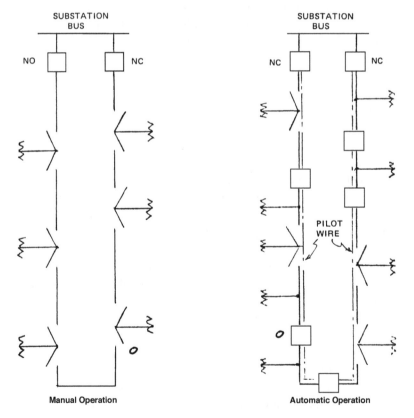

Figure 2-37(a). Primary Loop Systems

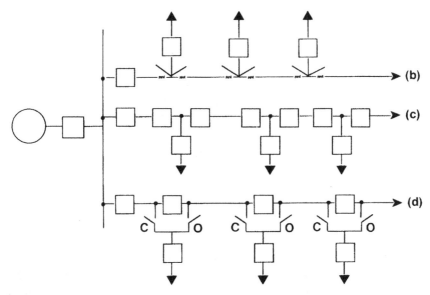

Figure 2-37(b). Radial type circuits showing several methods of sectionalizing.

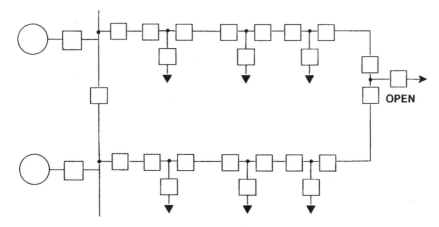

Figure 2-27(c). Loop type circuit showing methods of sectionalizing. One circuit breaker scheme as shown in Figure 2-37(b) also applies.

cading" one after the other, until the entire system is shut down (blackout). To obviate this occurrence, it is imperative that the circuit breakers nearest the fault open as fast as possible, clearing the fault from the circuit and stopping the rocking motion described above from continuing. This activity is often referred to as the "stability" of the system.

Primary Network Systems

In some instances where high reliability is desired, the primary network may prove less expensive than variations of the radial and loop systems. Similar to the secondary network, the primary mains of radial systems are connected together to form a network or grid. The grid is supplied from a number of power transformers or substations, supplied in turn at higher voltages from subtransmission or transmission lines. Circuit breakers between the transformer and grid are designed to open to protect the network from faults that may occur on the incoming high voltage lines, Figure 2-38. Faults on sections of the primary lines comprising the network are isolated by fuses and circuit breakers.

This type of system may be supplied from transformers located at conventional type substations and from smaller "unit" substations that are self-contained units including transformer, circuit breaker or breakers, relays, and other associated devices and equipment. They require less space and are strategically placed throughout the network, although obtaining sites in the midst of built-up areas may be more difficult. Moreover, some difficulty may be experienced in maintaining satisfactory operation of the voltage regulators on those interconnected feeders in the network on which regulators exist.

While the primary network has excellent voltage regulation and provides for load growth by simply adding substation capacity at or near the area of growth, proper operation of the network is difficult to attain. The loading of the feeders interconnected in the network must have sufficient reserve, sometimes not available, to enable the load to be carried by the remaining feeders when one (or more) becomes deenergized. Settings of relays associated with the feeder circuit breakers is also very difficult.

DISTRIBUTION SUBSTATIONS

Distribution substations serve as the source for primary distribution feeders. They receive bulk electric power at high voltages and reduce the voltage to distribution primary values. Associated with the transformation are provisions for protection from faults, from the elements and from overloads, for maintaining good voltage regulation, for obtaining data for monitoring the operations of the transmission and

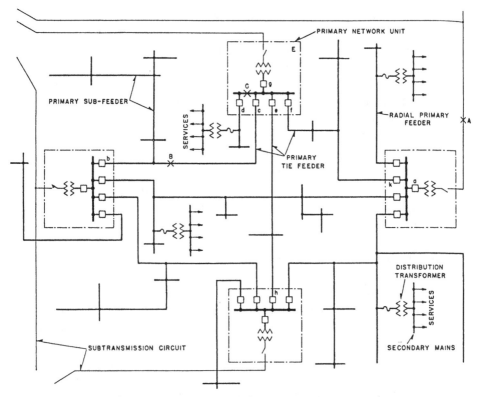

Figure 2-38. Typical Primary-Network Arrangement Using Breakers at Each End of Each Tie Feeder. *(Courtesy Westinghouse Electric Co.)*

distribution lines.

The location of a distribution substation is of great importance. While it should be situated as close to the load center as practical, this is not always readily obtainable for both economical and environmental reasons. The distance between substations, a product of site availability, is also an important consideration in the design of a distribution system. Both of these factors affect the size or capacity not only of the substation, but of the area load to be served. They influence largely the selection of the primary voltage of the circuits and, to a lesser extent, the type of primary and secondary mains employed in serving consumers. In some instances, they affect the decision concerning overhead vs. underground construction. In short, the location of the distribution substation is an integral part of the design of a distribution system.

Distribution substations may be of the outdoor type, completely enclosed in an indoor type, or a combination of the two. The final selection depending not only on economic factors, and future load growth, but on environmental, legal, and public relations factors.

The principal equipment generally installed in a distribution substation, together with auxiliary devices and apparatus are: power transformers, oil or air circuit breakers, voltage regulators, protective relays, air break and disconnecting switches, surge arresters, measuring instruments; in some instances, storage batteries, capacitors and street lighting equipment. This equipment is electrically and physically arranged for simplicity in construction and maintenance, including provisions for installing additional equipment to accommodate future load growth, with minimum disturbance of existing equipment. It is placed and interconnected in various arrangements by means of buses or cables to insure safety for workers and reliability of operation.

Power Transformers

Power transformers are larger in size than distribution transformers and may have auxiliary means for cooling. The latter may include fins on radiators attached to the tanks, fans blowing on the units, circulation of oil from the tank to external heat exchangers and back to the tank. Auxiliary tanks, mounted on top of the transformers, permit the oil to expand and contract without the hot oil making contact with the cool atmosphere, preventing condensation and sludge from contaminating the oil; these are sometimes referred to as conservators, Figure 2-39. The units may be single phase or three phase, and are provided with taps. While the line distribution transformers may be of either negative or positive polarity (although standards call for positive polarity), station power transformers are usually of negative polarity in accordance with ASA (American Standards Association) standards; the nameplate on the transformer usually identifies its polarity. The transformers may be connected in delta and Y combinations, similar to those described for line transformers.

Circuit Breakers

Circuit breakers are generally installed on both the incoming supply lines and the outgoing distribution feeders to protect the circuits in the event of fault, in addition to the job of energizing and deenergizing the circuits under normal conditions, including the maintenance of

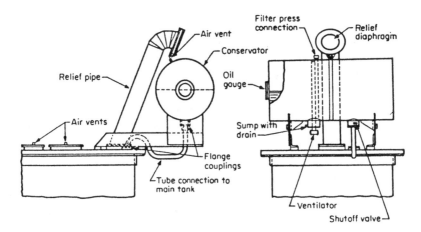

Figure 2-39. Diagram Showing Main Features of a Conservator (Courtesy General Electric Co.).

equipment and sectionalizing of lines. When a fault occurs on a feeder, extraordinary large currents will flow that the circuit breaker is called upon to interrupt. For example, assume a fault on a 7620-volt primary circuit just outside the substation, with a line impedance of (say) 0.2 ohms from the source to the fault. Then, by Ohm's Law, the current flowing would be:

$$I = \frac{E}{Z} = \frac{7620}{0.2} = 38100 \text{ amperes}$$

Such a large current will produce an alternating magnetic field about the conducting parts of the circuit breaker of such magnitude as to be destructive if allowed to flow for any relatively long period of time. Hence, the circuit breaker is designed to operate as quickly as possible and the arc that forms between opening contacts be quenched as rapidly as possible. In the first instance, the circuit breaker is made to open by the release of a spring; when the breaker is closed, by means of a coil (solenoid), the spring is wound and ready for the next operation. Both opening and closing are rapid operations consuming only a few cycles or less to open and several cycles or less to close.

Several schemes are employed to quench the arc, under oil, as quickly as possible; one employs the explosive action of the arc to blow itself out; another has the current in the arc produce a circulating magnetic field which also tends to blow out the arc quickly; still another, termed de-ion, has the arc travel over a series of closely spaced metal spacers, splitting the arc into a number of smaller ones more readily extinguished by the rapidly circulating magnetic field.

Air Circuit Breakers

Air circuit breakers may be used when fault currents are relatively small. They are of simple construction, low cost and low maintenance. Here, the insulating medium is air and easily ionized resulting in a severe and persistent arc. The current flows through coils creating a magnetic field that tends to force the arc from more rugged auxiliary contacts into ceramic chutes that stretch the arc, sometimes with the aid of compressed air, into extinction. Meanwhile, the main contacts open with little or no arcing.

Vacuum Circuit Breakers

Vacuum circuit breakers have the advantage of size and simplicity of construction, but maintenance is more complex and costly. Their interrupting rating, however, is higher than air circuit breakers but lower than oil circuit breakers. Here, the contacts open in a vacuum that, theoretically, cannot sustain an arc, but since no vacuum is perfect, a relatively small arc of short duration is produced; the heat generated is not easily dissipated.

Each of these types of circuit breakers may be single phase or three phase with all three phases contained in one tank but separated by partitions of material having high insulation value and laminated for greater strength. Special attention is given to the insulation of the circuit breaker parts, discussed under Insulation Coordination.

Voltage Regulators

Voltage regulators may be of two types, the induction regulator and the tap changing under load transformer (TCUL). Both types raise or lower the voltage from the substation power transformers, both work on the principle of the autotransformer.

The induction regulator has a primary coil connected across the circuit to be regulated and is wound on a steel core capable of rotating on the axis of the secondary. The secondary coil is stationary and is wound on a steel core; the cores of both coils constituting the magnetic circuit. The voltage induced in the secondary depends on the relative position of the two coils, adding when the primary coil rotates in one direction and subtracting when it rotates in the opposite direction. The reactance of the secondary coil causes a large voltage drop while the primary coil is rotating. A third coil, short circuited on itself, is mounted on the movable coil at right angles to the primary coil, acts to reduce this reactance. This type regulator, because of the difficulty with high voltage insulation on a moving coil, is generally limited to primary circuits of 5 kV and below. Figure 2-40.

The TCUL type changes the voltage in the outgoing primary feeder by varying the ratio of transformation, accomplished by means of taps

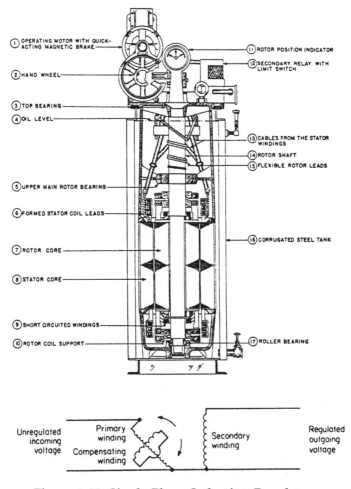

Figure 2-40. Single-Phase Induction Regulator
(Courtesy Westinghouse Electric Co.)

on the primary coil that can be changed while the autotransformer is energized, Figure 2-41. Small autotransformers connected between successive taps prevent the transformer from being disconnected from the line while taps are being changed. The regulated line is connected to the midpoint of these small autotransformers, the two ends of which are connected to the taps through "transfer switches." A voltage from one tap is added to or subtracted from the primary voltage with one of the transfer switches closed and the other open; another voltage is added to or subtracted from the primary voltage when both transfer switches are closed; a third voltage is added to or subtracted from the primary voltage when the second transfer switch is closed and the first open, Table 2-1. This type regulator is used when high voltage and relatively large loads are to be regulated.

A single-phase regulator adds or subtracts a voltage very nearly in phase with the line voltage, Figure 2-42a. This voltage changes in value with the position of the regulator, but does not change its phase relation.

A three-phase regulator adds or subtracts a voltage of constant value to the voltage across each phase, regardless of the position of the regulator. That voltage is not in phase with the line voltage, except at maximum and minimum, boost or buck, positions, Figure 2-42b. In the intermediate positions, the voltage increment is out of phase with the line voltage, except at 900 when no voltage is added or subtracted; that is, the system is in neutral. The three-phase increments rotate the line voltages out of their original phase positions. Care should be taken in paralleling three-phase regulators because of the out-of-phase resultant voltages; it may prove desirable to use two or three single phase regulators instead, Figure 2-42c.

Both types of regulators are controlled by means of a voltage sensitive relay, called a contact-making voltmeter, connected to the output side of the primary circuit. To control the voltage regulation at some point near the load center of the primary feeder, a "line-drop" compensator is introduced into the contact-making voltmeter circuit. The line-drop compensator contains a resister and reactor whose values represent in miniature the resistance reactance of the primary circuit to the point of regulation, Figure 2-43. The miniature values depend on the ratios of transformation of the instrument transformers on the primary circuit that connect to the contact-making voltmeter. Both the resistance and reactance elements may be varied in value.

Table 2-1. Sequence of Operation of TCUL Regulator Switches

Switch	Position								
	1	*2*	*3*	*4*	*5*	*6*	*7*	*8*	*9*
Transfer switch A	x x		x x x		x x x		x x x		x x
B		x x x		x x x		x x x		x x x	
C	x x x x x x x x x x								
Selector switch I	x x								
2		x x x							
3			x x x						
4				x x x					
5					x x x				
6						x x x			
7							x x x		
8								x x x	
9									x x

Courtesy Westinghouse Electric Co.

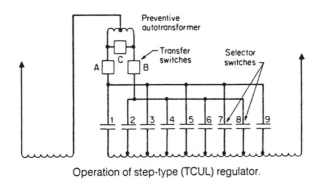

Operation of step-type (TCUL) regulator.

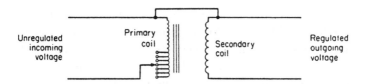

Figure 2-41. Step Type Regulator (TCUL)
(Courtesy Westinghouse Electrical Co.)

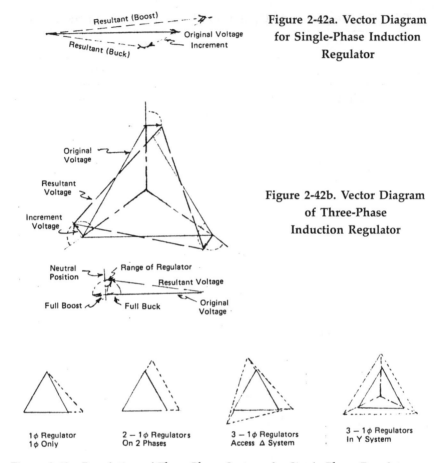

Figure 2-42a. Vector Diagram
for Single-Phase Induction
Regulator

Figure 2-42b. Vector Diagram
of Three-Phase
Induction Regulator

Figure 2-42c. Regulation of Three-Phase Systems by Single-Phase Regulators

Figure 2-42.
(Courtesy General Electric Co.)

Surge or Lightning Arresters

Surge arresters are installed on each conductor of the incoming supply feeders (subtransmission or transmission) and on those of the outgoing distribution feeders, as well as on other equipment that may be subject to voltage surges caused by lightning or switching. They usually consist of an air gap in series with some material having the characteristic of being an insulator at normal voltages but changing to a conductor when higher voltages are imposed on it, but returning to its

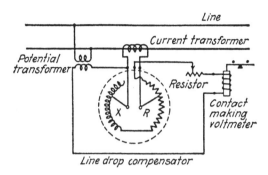

Figure 2-43. Schematic Diagram of Line-Drop Compensator and Contact-Making Voltmeter. *(Courtesy General Electric Co.)*

insulating condition when the high voltage is removed. The arresters are connected between the energized line or equipment and ground, and should be located as close as practical to the line or equipment to be protected. The ground resistance should be maintained as low as possible or the arrester will be ineffective.

A surge wave traveling over a conductor of a circuit may change its characteristics when it encounters a point of discontinuity, such as an open switch, a transformer, a change from overhead to underground, etc. At such points, the surge voltage wave may be reflected back on the conductor and the reflected voltage wave may add to or subtract from the original surge wave. The result is a crest voltage of double the original value or one that may tend to cancel the voltage of the surge wave, Figure 2-44.

In order for the arrester to fulfill its function to prevent excessive voltage stress from damaging the insulation of the protected line or equipment, it is essential that the characteristics of the insulation of the line or equipment be coordinated with the protective characteristics of the arrester. As there are many other devices for the protection of lines and equipment, such as circuit breakers, reclosers, fuses, etc., all of them must have their characteristics coordinated so that each of them may operate properly to provide protection, that is, protection of their insulation against failure.

Insulation failure may result from deterioration or puncture caused by prolonged overheating which, in turn, may depend on the duration and magnitude of the current flowing in the conductors. These may be greater than the design values of insulation, including factors of safety

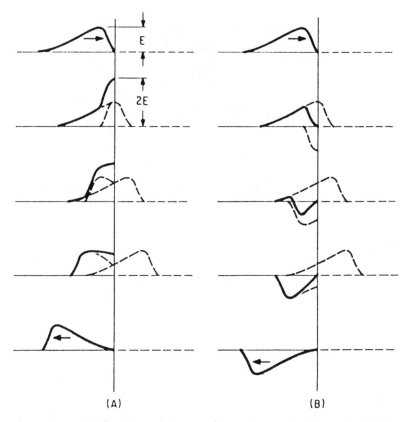

Figure 2-44. Reflection of Waves from Open Circuit End of Line
(Courtesy Westinghouse Electric Co.)

and manufacturer's tolerances. Lightning and switching surges cause very high heat to be generated by the high voltages and currents but of short duration.

Basic Insulation Level (BIL)

Insulation levels are designed to withstand surge voltages, rather than only normal operating voltages. Since the insulation lines and equipment is protected by arresters draining the surges rapidly before the insulation is damaged, the arrester must operate below the minimum insulation level that must withstand the surges. An example is shown in Figure 2-45a. The minimum level is known as the Basic Insulation Level (BIL) that must be that of all of the components of a system.

Insulation values above this level for the lines and equipment in the system must be so coordinated that specific protective devices operate satisfactorily below that minimum level.

In the design of lines and equipment considering the minimum level of insulation required, it is necessary to define surge voltage in terms of its peak value and return to lower values in terms of time or duration. Although the peak voltage may be considerably higher than normal voltage, the stress in the insulation may exist for only a very short period of time. For purposes of design, the voltage surge is defined as one that peaks in 1.5 microseconds and falls to one-half that value in 40 microseconds (thousandths of a second). It is referred to as a 1.5/40 wave, the steep rising portion is called the wave front and the receding portion the wave tail, Figure 2-46.

Insulation levels recommended for a number of voltage classes are listed in Table 2-2. As the operating voltages become higher, the effect of a surge voltage becomes less; hence, the ratio of the BIL to the voltage class decreases as the latter increases. Distribution class BIL is less than that for power class substation and transmission lines as well as consumers' equipment, so that should a surge result in failure, it will be on the utility's distribution system where interruptions to consumers are limited and the utility better equipped to handle such failures.

The line and equipment insulation characteristics must be at a higher voltage level than that at which the protecting arrester begins to spark over to ground, and a sufficient voltage difference between the two must exist. The characteristics of the several type arresters are shown in the curves of Figure 2-47. The impulse level of lines and equipment must be high enough for the arresters to provide protection but low enough to be economically practical.

Surges, on occasion, may damage the insulation of the protective device; hence, insulation coordination should include that of the protective devices. As there are a number of protective devices, mentioned earlier, each having characteristics of its own, the characteristics of all of these must be coordinated for proper operation and protection. These are discussed in Chapter 5.

Before leaving the subject of insulation coordination, such coordination also applies within a piece of equipment itself. The insulation associated with the several parts of the equipment must not only withstand the normal operating voltage, but also the higher surge voltage that may find its way into the equipment. So, while the insulation of the

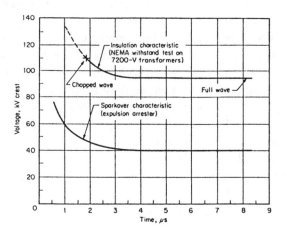

a. Insulation coordination *(Courtesy McGraw Edison Co.)*

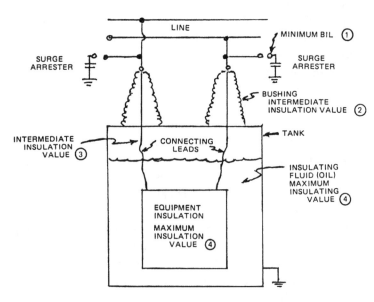

**b. Simplistic Diagram Illustrating Basic Insulation Level
and Insulation Coordination**

Figure 2-45

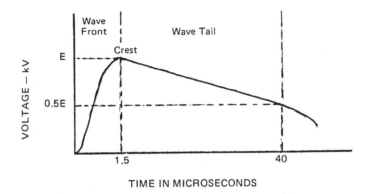

Figure 2-46. Surge Voltage 1.5 by 4.0 Wave

Table 2-2. Typical Basic Insulation Levels

Voltage class, kV	Basic insulation level, kV (standard 1.5- × 40-µs wave)	
	Distribution class	*Power class (station, transmission lines)*
1.2	30	45
2.5	45	60
5.0	60	75
8.7	75	95
15	95	110
23	110	150
34.5	150	200
46	200	250
69	250	350

*For current industry recommended values, refer to the latest revision of the National Electric Safety Code.

several parts is kept nearly equal, that of certain parts is deliberately made lower than others; usually this means the bushing. Since the bushing is usually protected by an air gap or arrester whose insulation under surge is lower than its own, flashover will occur across the bushing and the grounded tank. The weakest insulation should be weaker by a sufficient margin than that of the principal equipment it is protecting; such

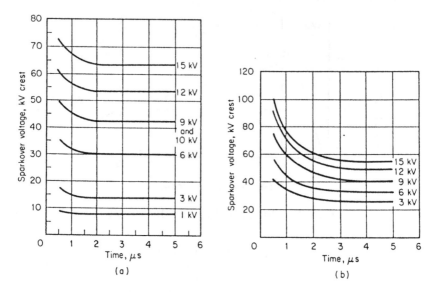

Figure 2-47. (a) Sparkover Characteristics of Distribution Value Arresters; (b) Sparkover Characteristics of Expulsion Arresters. (*Courtesy McGraw Edison Co.*)

coordinated arrangement restricts damage not only to the main parts of the equipment, but less so to parts more easily accessible for repair or replacement. The insulation of all parts of the equipment should exceed the basic insulation level (BIL). Figure 2-45b.

Other Substation Equipment

Other equipment that may be found in a distribution substation include switches, instrument transformers, measuring instruments, protective relays, and in some instances, capacitors, reactors, street lighting equipment, and storage batteries.

Switches

Switches may be of several types, the common characteristic is that none are designed to interrupt fault currents and the insulation or insulators associated with them must coordinate with the rest of the system and a BIL capable of withstanding voltage surges. In general, they consist of a conducting blade, hinged at one end, and a stationary contact on the other, both terminals mounted on suitable insulators that conform to the common insulator requirements of BIL coordination.

Almost every major line or equipment in a substation has associated with it a means of completely isolating it from other energized elements as a prudent means of insuring safety by preventing accidental energization. These simple switches, called disconnects, or disconnecting switches, are usually installed on both sides of the equipment or line upon which work is to be done. They should not be operated while the circuit in which they are connected is energized, but only after the circuit is deenergized. As a further precaution, they may be opened by means of an insulated stick that helps the operator keep a distance from the switch. Locking devices are sometimes provided to keep the disconnects from being opened accidentally or from being blown open during periods of heavy fault currents passing through them. Although not designed to be closed to energize the line or equipment with which they are associated, in certain circumstances they may be closed, using special care to close them firmly and rapidly. Disconnects may be single-blade units or multiple units operated together.

Air break switches have characteristics similar to disconnects, but have the stationary contacts equipped with arc suppressing devices that enable them to be opened while energized, but recognizing a limitation as to the current that may be safely interrupted. The device may be a simple arcing horn which stretches the arc that may form until it cannot sustain itself. Another type has a flexible "whip" attached to the stationary contact that continues the contact with the moving blade until a point is reached at which the whip snaps open very rapidly extinguishing any arc that may form. Still another has an interrupting unit mounted at the free end of the switch blade and which suppresses the arc within the unit as the switch opens; the interrupting unit may contain a vacuum chamber, or a series of grids which cause the arc to break up into smaller ones and are more readily extinguished. Both types function similarly as larger units do in circuit breakers.

Oil switches have the blade open from its contact under oil which suppresses any arc that may form. They have higher current breaking capacity than air break switches and are particularly suited for underground systems or where moisture or pollution make air switches impractical. However, they are more expensive at first cost, to operate and to maintain.

Instrument Transformers

Instrument transformers are used in substations because of the impracticability of measuring values of current and voltage with ordinary meters, not only because they may be beyond the range of these instruments, but because the insulating problems make their direct use impractical. Instrument transformers act in the same way as distribution transformers, but have a much greater accuracy in their ratios of transformation. Both current transformers (CT) and potential transformers (PT) are insulated to withstand the voltage of the circuits with which they are associated, and also conform to the BIL and insulation coordination of the systems of which they are a part.

Current transformers usually have a secondary rating of 5 amperes and potential transformers a secondary rating of 150 or 300 volts, and sometimes ratings of 250 and 500 volts. Both instrument transformers also isolate the meters and relay circuits from the high voltages of the incoming and outgoing feeders. They are rated in volt-amperes, and the load they carry is referred to as their burden. The turn ratios of the current transformer usually results in stepping up the voltage in its secondary. This voltage is reduced considerably by the impedance of the instrument or device connected across the terminals of the secondary; if the secondary is left open with nothing connected to it, the voltage developed may be so high as to be unsafe and the secondary of the current transformer is provided with a short circuit device when opened. Current transformers are also provided with polarity marks to aid in their proper connection when paralleled with other current transformers.

Relays and Meters

The operation of protective relays is discussed in a separate chapter on protection. It is assumed the reader is familiar with the operation of meters that indicate and record data desired for observing the operation of substations, including both incoming and outgoing feeders.

Capacitors

Capacitors for regulating bus voltage by correction of power factor may also be installed at distribution substations. This mode of operation has been described earlier. Provision is made for switching some of the units in the capacitor bank off and on as required to maintain voltage regulation. Capacitors are connected to the bus through fuses that serve to clear a faulted capacitor unit so that other energized elements in the substation are not affected.

Reactors

Reactors may be connected in series with the line or equipment to limit the flow of fault current that may flow through them. They may also be connected in series with a transformer that may be paralleled with one of dissimilar characteristics in order to obtain an equitable balance of loads between them. Where distribution feeders are cables with metallic sheaths and operate at relatively high voltages, reactors may be installed to counter some of the cable's capacitance effect.

Street Lighting Equipment

Street lighting equipment supplying series street may sometimes be found in distribution substations. Details of equipment and mode of operation are contained in Appendix E.

Storage Batteries

Storage batteries may be found in some distribution substations, usually the larger and more important ones. Ordinarily, circuit breakers, indicating lamps, and other devices are operated by alternating current from a transformer at the substation assigned to this purpose. When operation of these devices, including the reclosing of circuit breaker mechanisms, is of special importance, they are supplied from a direct current source from a bank of storage batteries. Operation may be at nominal voltages of 6, 12, 24, 48 or 120 volts, although 24, 48 and 120 volt systems are preferred as possible voltage drop from poor connections will still leave sufficient voltage for proper operation of relays, breakers, and auxiliaries.

The batteries are kept charged continuously, being connected to a 120-volt alternating-current supply through rectifiers. Lead cell batteries must be ventilated as they give off hydrogen and oxygen gases, a potentially explosive mixture. Batteries are rated in ampere-hours.

Fuses

Fuses provide protection by the melting of a fusible link when current exceeding their rating flows through them, Figure 2-48a, b, c. They come in a variety of types and ratings for both voltage and current values. From the relatively low voltage line fuse associated with distribution transformers to those on high voltage transmission lines, they all operate on the same principle and each has its own time-clearing characteristic which must be considered with the characteristics of other

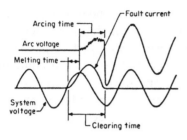

a. Oscillogram of link melting and fault current interruption.

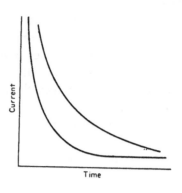

b. Typical time-current characteristic for fuses.

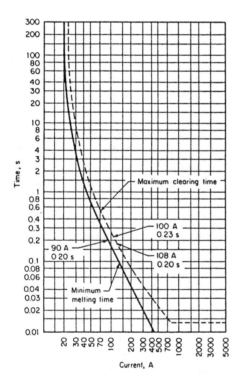

c. Typical time-current characteristic for a 10-K fuse link.

Figure 2-48
(Courtesy McGraw Edison Co.)

protective devices on the system (see Chapter 5). The insulation of their mountings; must also subscribe to the BIL and coordinate with the system insulation.

Distribution Substation Bus Arrangement

Bus arrangements provide for varying degrees of reliability on both input and output sides. On the incoming subtransmission or transmission supply feeders, the arrangements may vary from one feeder to a multiplicity of feeders and the disconnecting facilities from an air-break switch to a multiplicity of more expensive circuit breakers, Figure 2-49. Each additional circuit breaker provides greater flexibility in arranging the supply feeders so that a higher degree of reliability can be maintained should one (or more) of the supply feeders be out of service from fault condition or for maintenance. The same advantages also exist for similar conditions on the high voltage bus arrangement.

On the low voltage outgoing primary bus arrangement, a similar variation from an air-break switch to a multiplicity of circuit breakers provide flexibility in maintaining the bus energized during fault or maintenance occurrences; again, the greater the number of circuit breakers, the greater the cost.

Another variation in the arrangements associated with individual distribution primary feeders employs a varying number of circuit breakers, Figure 2-50. Even greater reliability may be achieved through the flexibility of arrangements by a combination of all or some of the buses described.

The insulation of all of the buses, switches, circuit breakers, etc., of the several arrangements must also conform to BIL and coordination requirements described earlier.

Conversion of Primary Circuits to Higher Voltages

Load density increases in particular areas are often taken care of by increasing the voltage of the existing primary circuit. This frequently involves changing the delta connection to a Y connection of lines and transformers at both the substation and on the feeder (e.g. 2400 delta to 4160Y; 7620 delta to 13200Y; etc.). In such instances, practically all of the facilities, including poles, conductors, insulators, transformers, cutouts, surge arresters, are adequate to operate at the higher nominal voltage. If a secondary neutral conductor does not exist, it should be installed to act as a common ground or neutral for the higher Y voltage circuit.

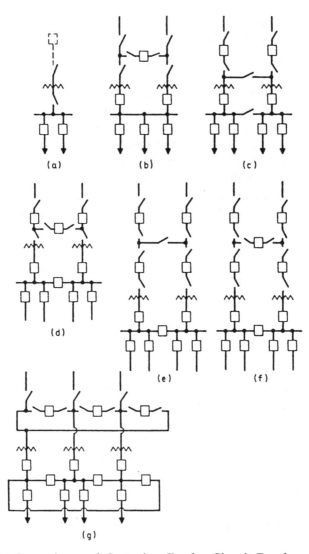

Figure 2-49. Incoming and Outgoing Feeder Circuit Breaker Arrangements. *(Courtesy' Westinghouse Electric Co.)*

The changeover is accomplished principally with the aid of one or more transformers having an input of the lower voltage and an output at the higher voltage. These can be single-phase or three-phase units. They can be pole-mounted or mobile, sometimes mounted on a platform on a truck. The general method is to use these transformers to pick up

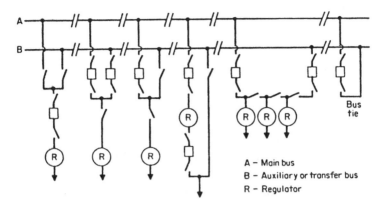

Figure 2-50(a). Arrangement of Distribution Feeder Buses at Substations. *(Courtesy Westinghouse Electric Co.)*

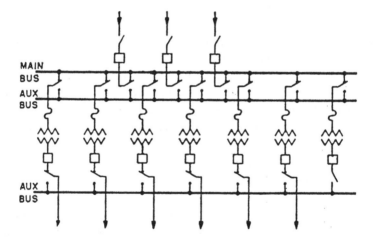

Figure 2-50(b). Distribution substation with high-voltage double bus and low-voltage auxiliary bus and individual transformation in each primary feeder resulting in relatively low interrupting duty on feeder breakers.

small sections of the feeder at a time, the three-phase units picking up part of the feeder main or trunk and the single-phase the laterals. The details depend on: the amount of load limited by the capacity of the transformers; the practical points of connection, preferably at sectionalizing points utilizing the switches at these points; the availability of adjacent feeders to pick up parts of the circuit being worked on;

the availability of manpower to determine how much work of reconnection may be done during periods of light load; in the substation, whether the load of the feeders on one bank can be picked by other available banks, allowing time for one bank to have its connections changed from delta to Y on the outgoing primary supply. The timing of the reconnection of the transformers at tile substation depends on whether the field conversion is to start at the near end or the far end of the feeder, or the laterals of the feeder under consideration.

In some instances, after a portion of the circuit has been converted, it may be practical and economical to leave the "temporary" conversion transformers in place on the feeder for a longer "permanent" period of time.

In general, whatever outages that may be necessary to permit reconnection of line transformers, laterals, etc., to be done safely, should be kept to a minimum.

When conversion to a higher voltage is not based on reconnection of delta to Y circuits, utilizing most of the facilities in place, much preparatory work will need to be done. Generally, existing poles and conductors will be adequate. Insulators may need to be changed, higher primary voltage transformers and their accessories will need to be replaced (fused cutouts and arresters), new units installed a pole or one span away from the existing facilities. When all of the preparatory work is completed, the conversion in the field will follow the same general procedure as described above.

Replacement of transformers, circuit breakers, and revamping of buses at the substation may require building of temporary by-passes while the work is being done; employment of a mobile transformer or substation may simplify and expedite the work. In some instances, where the incoming transmission supply voltage is also being changed, the work may become more involved; more temporary work may be necessary and, again, a mobile substation may prove helpful.

When the conversion involves changes to a relatively higher voltage (say) 2400/4160-volt system to a 7620/13200-volt distribution system, close attention should be given to make sure of proper clearances between conductors, between conductors and nearby structures, adequate tree trimming, etc., based on the requirements of the NESC as a minimum, and other safety and local regulations. For the higher voltage, more outages of greater duration, or recourse to live-line procedures, or both, may be necessary, as well as the erection of temporary bypasses.

While these add to the cost of the conversion, safety should never be sacrificed; these additional costs should be included in the economic studies comparing this method of serving increases in area load densities against other alternatives, such as adding additional substations.

Such conversions to higher distribution voltages for underground systems rarely follow the methods described above for overhead systems. Generally, it is necessary to provide for either a complete reconstruction, essentially installing a new system, or the addition of new facilities at existing voltages, but changing radial systems to networks.

Parenthetically, similar methods may be used in the conversion of two-phase circuits to three-phase operation. In place of the transformers that change one primary voltage to another, transformers connected in the Scott connection or Y to three-phase connection are employed. Individual two-phase loads may still be served by similar phase transformation connected small transformers with secondary voltage values.

Chapter 3

Subtransmission System Electrical Design

Subtransmission lines serve as incoming supply lines to distribution substations. They usually operate at nominal voltages of 23, 45, 69 and 138 kV, voltages between distribution and transmission line values, although some systems employ the same distribution as well. Subtransmission circuits may be arranged in a simple radial pattern, or in open or closed loops that may, in some instances, emanate from or act as ties between two or more bulk power sources. Both radial and loop systems exhibit somewhat the same characteristics as primary distribution circuits. Where a distribution substation is supplied by two or more subtransmission feeders or two or more sections of a loop feeder, the several feeders or sections of feeders may be routed along separate rights-of-way for greater reliability. Figures 3-1, 3-2, 3-3.

Subtransmission feeders may also be connected in a grid or network manner, but the number of circuit breakers involved and the complexity of protective relay schemes limit their use to relatively few areas where a very high degree of reliability may not be achieved more economically with other types of systems. Where the grid subtransmission system interconnects two or more power sources, greater reliability may be achieved, but system protective relay problems become even more complex. Figure 3-4.

SUBTRANSMISSION SUBSTATIONS

Subtransmission substations typically supply several distribution substations. Where subtransmission feeders are ties between two or more subtransmission substations, they may also serve to equalize loads on the incoming supply transmission feeders. Because of the considerably greater loads carried by subtransmission substations, continuity of

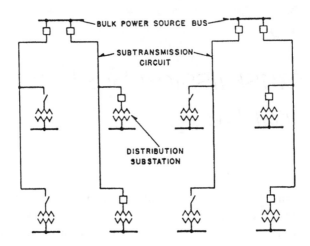

Figure 3-1. Simple Form of Radial Type Subtransmission Circuit.
(Courtesy Westinghouse Electric Co.)

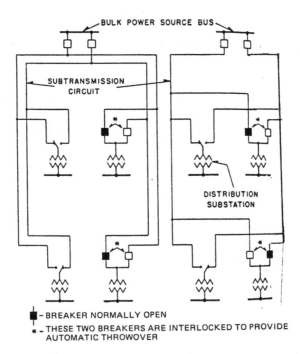

Figure 3-2. Improved Form of Radial Type Subtransmission Circuits
(Courtesy Westinghouse Electric Co.)

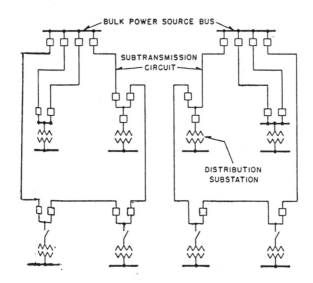

Figure 3-3. A Parallel or Loop Circuit Subtransmission Layout
(Courtesy Westinghouse Electric Co.)

service receives greater consideration in their designs than for distribution substations. In general, subtransmission substation bus and equipment arrangements are similar to those found in the distribution substation designed for greater service reliability: buses sectionalized, main and transfer buses, ring buses, and a greater use of circuit breakers for ties and load transfers.

Some of the subtransmission substation arrangements, are shown in the one-line diagrams of Figure 3-5, and refer to the busing and switching at the subtransmission voltage level. Starting with the simplest arrangement, the arrangements become more complex as the degree of continuity and reliability become greater. The possible exception is the ring bus arrangement which requires only one circuit breaker for incoming or outgoing circuits; moreover, each outgoing circuit has, in effect, two sources of supply. With fewer circuit breakers than other arrangements, it provides relatively good service reliability; any circuit breaker in the ring may be taken out of service for maintenance or other work without disrupting the remainder of the system. It must be remembered that the normal load capacity and the short circuit duty imposed on the circuit breakers in subtransmission substations are higher than those imposed on those in distribution substations and are more

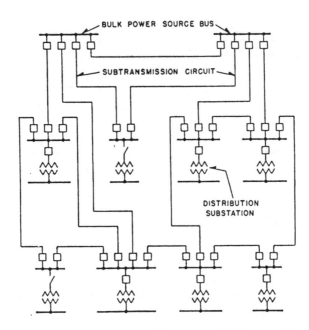

Figure 3-4. Network or Grid Form of Subtransmission
(Courtesy Westinghouse Electric Co.)

costly. The higher voltages involve higher insulation requirements with
their separate BIL and coordination requirements. Hence, the number of
circuit breakers involved in any arrangement should be held to as few
as is consistent with the continuity and reliability of service desired.

SUBTRANSMISSION SYSTEM CAPABILITY

In many subtransmission circuit arrangements, power flow from
the subtransmission substation or substations to the distribution substations may be from two or more directions. Subtransmission systems
usually have more flexibility than transmission systems. Transmission
lines are designed to meet power flow conditions between generating
stations, system interchange points, or transmission substations.
Subtransmission systems are designed to provide supply to one or more
distribution substations and for possible additions.

The capability requirements are based on the ultimate number, the
kVA rating and distance between distribution substations. The kVA rat-

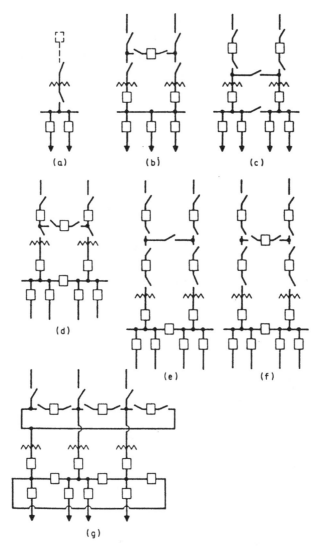

Figure 3-5. Incoming Feeder Circuit Breaker Arrangements
(Courtesy Westinghouse Electric Co.)

ing of a distribution substation, in turn, is based on the area it serves, involving load density, and the voltage, number and allowable loading of the primary feeder. As the subtransmission feeders are electrically (and physically) interconnected with the distribution substation, the same factors should be taken into account in the planning and design of subtransmission feeders. Various combinations of these factors are investigated to determine the optimum economic design of the various components of a subtransmission system, including limits set by permissible 1^2R losses and voltage drop.

Factors that affect the cost of subtransmission systems include:

1. Arrangement of subtransmission circuits (radial, loop, grid)

2. Arrangement of distribution substations (radial, primary network, low voltage secondary network, duplicate service, etc.)

3. Load density

4. Distribution primary feeder voltage

5. Distribution substation rating

6. Subtransmission voltage

7. Subtransmission substation rating

Transmission voltage influences the cost of subtransmission substation transformers, high voltage busing and switching.

The cost analysis is made of various combinations of these factors to determine the most economical design. Chapters 2 and 4 include data on the several elements of the subtransmission system that are common to the distribution and transmission systems.

Chapter 4

Transmission System Electrical Design

The functions of a transmission system are:

1. To transport electric power from a generating source to a central point (a transmission substation) from which it may be transported to other central points, Figure 4-1.

2. To transport power in bulk quantities from a central point (transmission substation) to wholesale delivery points (subtransmission substations).

3. To act as tie points with interconnecting transmission lines from other power systems for emergency or economic reasons.

SELECTION OF VOLTAGE

Like the selection of subtransmission and distribution voltages, the selection of transmission voltages generally follow the same procedures. Several likely standard voltages are studied and the cost of losses and the carrying charges on the overall investment for each are compared; those approximating each other most closely determine, in conformity with Kelvin's Law, the selection. Included in the study, therefore, should be some voltage that, under normal circumstances, may appear uneconomical, but whose consideration of capacitors to increase the line capacity would increase the economic limit of power transmission. At higher voltages, however, the increase in reactance of the terminal transformers tend to offset the gain obtained from the capacitors.

Other factors, as in the case of subtransmission and distribution systems, may outweigh the economic selection; e.g., the desirability of

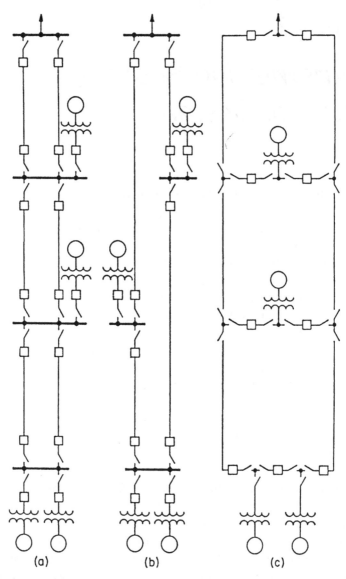

| (a) | (b) | (c) |

Figure 4-1(a). Fundamental Schemes of Transmission (a) Fully-Sectionalized Supply, (b) Looped-in Supply, (c) Bussed Supply. (*Courtesy Westinghouse Electric Co.*)

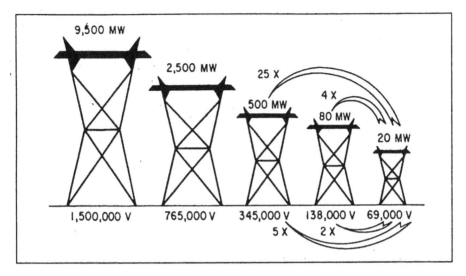

Figure 4-1(b) Comparison of power-carrying capabilities and operating voltages.

Standard Transmission Voltages

13,800	23,000	34,500	46,000	69,000	115,000	138,000
230,000	354,000	500,000	765,000	1,000,000	1,500,000	

interconnecting with other systems in a pool or grid, would give preference to the voltage common to the grid.

SELECTION OF CONDUCTOR

Closely associated with the choice of voltage is the choice of conductor which, in turn, is affected by the spacing of the conductors as well as the effects of lightning and switching surges. Depending on the reliability sought, some additional spacing may be considered beyond that required for normal voltage requirements. In determining the reactance of a transmission line, the "equivalent spacing" as well as the number of suspension insulators used in a string are factors to be considered. Equivalent spacing is that spacing that would give the same inductance and capacitance values as if an equilateral triangle arrangement of conductors is used. It is usually impractical to use the equilateral conductor arrangement for design purposes. Equivalent spacing may be obtained from the formula:

$$D = \sqrt[3]{ABC}$$

where A, B, and C are the actual distances between conductors.

In general, the same observations apply to cables. The equivalent spacing is referred to as the Geometric Mean Distance (GMD). Cables come in so many sizes, types, insulation, conductor shapes, sheaths, etc., that for reactance and capacitance factors, constants, etc., reference should be made to the manufacturer's data.

REGULATION AND LOSSES IN
A TRANSMISSION LINE

Ordinarily, the conductor size required to transmit a given current varies inversely as the square of the voltage. The saving in conductor size for a given loss becomes less as the voltage becomes higher. This is because of greater leakage over insulators and corona (energy escaping from the conductor) losses become significant. In addition to these two losses, the charging current, which increases as the transmission voltage goes higher, may either increase or decrease the current in the circuit, depending on the power factor of the load current and the relative amount of the leading and lagging components of the current in the circuit. Any change in the current of the circuit will therefore be accompanied by a corresponding change in the I^2R loss. Indeed, these sources of additional loss may, in some cases of long circuit extensive systems, contribute materially toward limiting the transmission voltage.

Because of the greater capacitance effect, voltage regulation and line losses for short transmission lines and long ones will be different. In the analysis that follows, all line voltages are line to neutral voltages unless denoted by the subscript L. For line to line voltage, the impedance drop should be multiplied by 2 for single-phase lines or by $\sqrt{3}$ for three-phase lines.

For practical purposes, in transmission lines of some 30 miles in length and of voltages under 40 kV, the capacitance effect can be safely neglected. For longer lines, the distributed capacitance and its charging current assumes greater importance. No definite length, however, can be assigned as the dividing point between short and long transmission lines.

Where capacitance can be neglected, a transmission line can be viewed as a concentrated impedance:

$$Z = R + jX$$

$$\text{or } 2s = rs + jxs$$

where s = series impedance of one conductor, in ohms per mile
       r = resistance of one conductor, in ohms per mile
       x = inductive reactance of one conductor, in ohms per mile
       s = length of line, in miles

a

The equivalent single-phase current is given in Figure 4-2, together with the vector diagram showing the relation between the line current and line to neutral voltage at both ends of the line. This relationship in analytic terms is indicated by the equation:

$$E_S = E_R + IZ$$

The following symbols are used in the accompanying discussion

E and I are vector quantities
$\bar{E}$ and $\bar{I}$ are absolute magnitudes of the quantity
$\hat{E}$ and $\hat{I}$ are conjugates of the vector quantities.

In the analyses of "long" transmission lines, it is necessary to consider that the charging current of the line varies directly with the voltage of the line and inversely with the load current. Such a line can be considered as an infinite number of series impedances and shunt capacitors connected as shown in Figure 4-3. The current I_R is unequal to I_S in both magnitude and phase position because some current is shunted through the capacitance between phase and neutral. The relationship between E_S and E_R for a long line is different from that for a short line because of

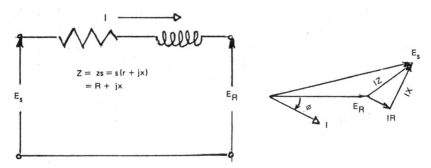

Figure 4-2. Equivalent Transmission Circuit—Line to Neutral—For Short Lines

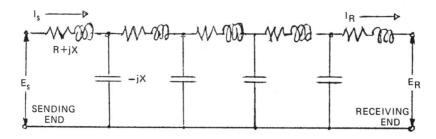

Figure 4-3. Equivalent Transmission Circuit —Line to Neutral—For Long Line

the progressive change in the current due to the shunt capacitance.

If E_S and E_R are the phase to neutral voltages and I_S and I_R are the phase currents, the equations relating the sending end voltages and currents to the receiving end quantities are:

$$E_S = E_R\left[\cosh\left(S\sqrt{\frac{2}{Z^t}}\right)\right] + I_R\left[\sqrt{2Z^t}\sinh\left(S\sqrt{\frac{2}{Z^t}}\right)\right]$$

$$I_S = E_R\left[\frac{1}{\sqrt{2Z^t}}\sinh\left(S\sqrt{\frac{2}{Z^t}}\right)\right] + I_R\left[\cosh\left(S\sqrt{\frac{2}{Z^t}}\right)\right]$$

where Z^t is the shunt impedance of the line in ohms per mile
          X^t is the capacitance reactance in megohms per mile

These equations can be more conveniently written in terms of the so-called ABCD constants:

$$E_S = AE_R + BI_R \quad \text{and} \quad I_S = CE_R + DI_R$$
$$E_R = AE_S + BI_S \quad \text{and} \quad I_R = -CE_S + DI_S$$

where the transmission line is symmetrical, $D = A$ and $AD - BC = 1$.

The Equivalent π Circuit

Referring to Figure 4-4, the equivalent impedance Z'_{eq}:

$$I_{R'} = \frac{E_R}{Z'_{eq}} \qquad I_{S'} = \frac{E_S}{Z'_{eq}}$$

$$E_S = E_R\left(1 + \frac{Z_{eq}}{Z'_{eq}}\right) + I_R Z_{eq}$$

$$I_S = E_R\left(\frac{2}{Z'_{eq}} + \frac{Z_{eq}}{Z'^2_{eq}}\right) + I_R\left(1 + \frac{Z_{eq}}{Z'_{eq}}\right)$$

Equating

$$E_S = E_R\left(1 + \frac{Z_{eq}}{Z'_{eq}}\right) + I_R Z_{eq} = AE_R + BI_R$$

$$Z_{eq} = B \qquad \text{and} \qquad A = 1 + \frac{Z_{eq}}{Z'_{eq}}$$

from which $Z'_{eq} = \dfrac{B}{A - 1}$.

The Equivalent T Circuit

Referring to Figure 4 – 5, the equivalent impedances are:

$$Z_T = \frac{A - 1}{C} \qquad \text{and} \qquad Z'_T = \frac{1}{C}$$

Voltage Regulation on Short Lines'(from Receiver)

In the following equation, the size of the power factor angle ϕ depends on whether the current is lagging or leading. For lagging power factor ϕ and $\sin\phi$ are negative; for a leading power factor, ϕ and $\sin\phi$ are positive. The $\cos\phi$ is positive for either lagging or leading current. Figure 4-6a.

$$E_R = \bar{E}_R = \text{reference}$$

$$I = \check{I}\cos\phi_R + j\check{I}\sin\phi_R$$

$$Z = R + jX = rs + jxs$$

$$E_s = E_R + IZ$$

or $E_S = (\bar{E}_R + \check{I}R\cos\phi_R - IX\sin\phi_R) + j(\check{I}X\cos\phi_R + \check{I}R\sin\phi_R)$. If the $\check{I}R$ and $\check{I}Z$ drops are relatively small (say 10%) of $\bar{E}_R$, E_S can be determined for normal power factors (say 80% ±) by neglecting its quadrature component:

$$\bar{E}_S = \bar{E}_R + \check{I}R\cos\phi_R - \check{I}X\sin\phi_R$$

and the voltage regulation of a line is usually the percent drop with reference to $\bar{E}_R$.

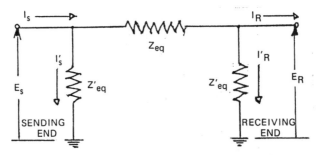

Figure 4-4. Equivalent 7r Circuit - Long Transinission Lines

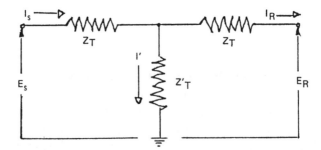

Figure 4-5. Equivalent T Circuit - Long Transmission Lines

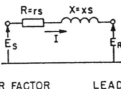

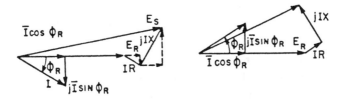

(a) FOR KNOWN RECEIVING END CONDITIONS

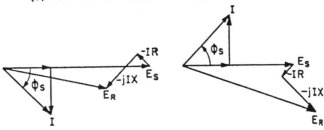

(b) FOR KNOWN SENDING END CONDITIONS

Figure 4-6. Vector Diagram for Determining Voltage Regulation of Short Lines. *(Courtesy Westinghouse Electric Co.)*

$$\% \text{ Regulation} = \frac{\bar{E}_S - \bar{E}_R}{\bar{E}_R} \, 100$$

$$= \frac{\bar{I}_S}{\bar{E}_R} \left(r \cos\phi_R - x \sin\phi_R \right) 100$$

at receiving end kVA $= \dfrac{3\bar{E}_R \bar{I}}{1000} = \dfrac{3\bar{E}_L I}{1000}$ where $\bar{E}_L$ is the line voltage at the receiving end; then

$$\% \text{ Regulation} = \frac{1000 \text{ kVAs}}{E_L^{\,2}} \left(r \cos\phi_R - x \sin\phi_R \right) 100$$

Using the regulation calculated from these equations, the receiving end voltage determined will be reasonably accurate, provided the resistance and reactance drops are not excessive (say less than 10%). The percent variation from its own correct value, however, may be great depending on its actual magnitude, hence such equations are not sufficiently accurate for determining load limits for fixed voltage regulation. It will also be observed that the amount of load that can be transmitted over a given line at a fixed regulation varies inversely with the load.

Voltage Regulation for Short Line (from Sending End)

From known sending end conditions, the receiving end voltage, ES, is used as the reference vector as shown in Figure 4-6b:

$$E_S = \bar{E}_S = \text{reference}$$
$$E_R = E_S - IZ$$

and $\bar{E}_R = \left(\bar{E}_S - \hat{I}R\cos\phi_S + \hat{I}X\sin\phi_S\right) - j\left(\hat{I}X\cos\phi_S + \hat{I}R\sin\phi_S\right)$

Neglecting the quadrature component of E_R:

$$\bar{E}_R = \bar{E}_S - \hat{I}R\cos\phi_S + \hat{I}X\sin\phi_?$$

Resistance Losses of Short Line

Total I^2R loss of a three-phase line is three times the product of the total resistance of one conductor and the square of its current

$$\text{Loss} = 3\hat{I}^2R \text{ (watts)}$$

in percent of desired kW load

$$\% \text{ Loss} = \frac{1.73\hat{I}rs}{\bar{E}_L\cos\phi_R} 100$$

where it is desired to determine the amount of power that can be delivered without exceeding a given percent loss:

$$kW = \frac{\bar{E}_L^{\,2}\cos^2\phi R}{1000rs} \times \frac{(\% \text{ Loss})}{100}$$

It may be observed that the amount of power that can be transmitted for a given percent loss varies inversely with the length of line and directly with the loss.

Regulation of Long Lines (from Receiving End)

The effect of charging current on the regulation of transmission lines may be determined from the equivalent π circuit; the vector diagrams for the known load conditions are shown in Figure 4-7a. The voltage drop in the series impedance Z_{eq} is produced by the load current I_R plus the charging current E_R/Z'_{eq} flowing through the shunt impedance at the receiver end of the line. For a given line, this latter current is dependent only on the receiver voltage E_R.

In considering the charging current, one method is to determine first the net current

$$I'_{eq} = I_R + \frac{E_R}{Z'_{eq}}$$

that flows through Z_{eq} together with its power factor angle ϕ_{eq}. Using the equivalent series impedance Z_{eq} and the current instead of the load current, all of the analytic expressions developed for short lines are applicable:

$$I'_{eq} = I_R + \frac{E_R}{Z'_{eq}}$$

$$= \bar{I}_R \cos\phi_R + j\left(\frac{\bar{E}_R}{Z'_{eq}} \pm \bar{I}_R \sin\phi_R\right)$$

$$= \bar{I}_{eq}\cos\phi_{eq} \pm j\bar{I}_{eq}\sin\phi_{eq}$$

$$E_S = E_R + I_{eq}Z_{eq}$$

The equivalent terminal conditions are shown in Figure 4-7a.

Regulation of Long Lines (from Sending End)

In this instance, the equivalent current flowing through Z'_{eq} may be determined as the difference between I_S and I'_S, the current in the shunt reactance at the sending end of the equivalent circuit:

$$I_{eq} = I_S - \frac{E_S}{Z'_{eq}}$$

$$= \bar{I}_S \cos\phi_S + j\left(\frac{\bar{E}_S}{\bar{Z}'_{eq}} \pm \bar{I}_S \sin\phi_S\right)$$

$$= \bar{I}_{eq} \cos\phi_{eq} \pm j\bar{I}_{eq} \sin\phi_{eq}$$

$$E_R = E_S - I_{eq} E_{eq}$$

The vector diagrams are shown in Figure 4-7b.

Resistance Losses on Long Lines

The effect of charging current on line losses can be treated as it was for Regulation of Long Lines (from Receiver End) above. Referring to Figure 4-7, the losses can be considered to be due to the current I_S:

$$I_{eq} = I_R + I'_R = I_S - I'_S$$

flowing through the equivalent resistance, R_{eq}.

Thus in terms of load current:

Loss for lagging power factor

$$= 3R_{eq}\,(I_R + I'_R)^2 \text{ watts}$$

$$= 3R_{eq}\left(\bar{I}_R^2 - \frac{2\bar{I}_R E_R}{\bar{Z}'_{eq}}\sin\phi_R + \frac{\bar{E}_R^2}{\bar{Z}'^2_{eq}}\right) \text{ watts}$$

Loss for leading power factor:

$$= 3R_{eq}\left(I_R^2 - \frac{2\bar{I}_R \bar{E}_R}{\bar{Z}'_{eq}}\sin\phi_R + \frac{\bar{E}_R^2}{\bar{Z}'^2_{eq}}\right) \text{ watts}$$

Circle and Loss Diagrams

Equations for line currents, power, and resistance losses can be expressed as functions of the terminal voltages and system constants.

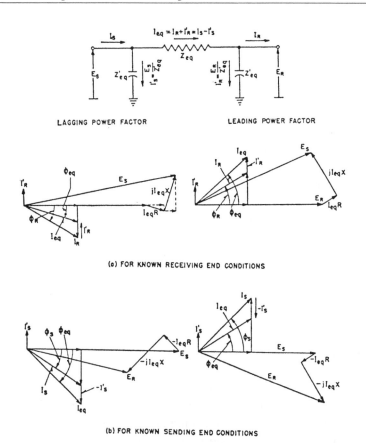

Figure 4-7. Vector Diagrams for Determining Voltage Regulation of Long Lines. *(Courtesy Westinghouse Electric Co.)*

Such equations and graphical representation of these are found convenient for the more common types of performance problems. The graphic form of the power and current equations are very similar and are known as "circle diagrams."

Vector Equations for Power

The vector expressions for power, the product of the current and the conjugate of the voltage may be written:

$$P + jQ = \bar{E}\hat{I}$$

and from the vector diagram of Figure 4-8

$$E = \bar{E}\cos\theta_e + j\bar{E}\sin\theta_e$$

$$\bar{E} = E\cos\theta_e - jE\sin\theta_e$$

$$I = \hat{I}\cos\theta_i + j\hat{I}\sin\theta_i$$

$$\hat{E}I = \bar{E}(\cos\theta_e - j\sin\theta_e)\hat{I}(\cos\theta_i + j\sin\theta_i)$$

$$= \bar{E}\hat{I}(\cos\theta_e\cos\theta_i + \sin\theta_e\sin\theta_i) + j(\cos\theta_e\sin\theta_i - \sin\theta_e\cos\theta_i)$$

Since $\cos(\theta_e - \theta_i) = \cos\theta_e\cos\theta_i + \sin\theta_e\sin\theta_i$
and $\sin(\theta_e - \theta_i) = \sin\theta_e\cos\theta_i - \cos\theta_e\sin\theta_i$
then $E\hat{I} = \bar{E}\hat{I}\cos(\theta_e - \theta_i) + j\bar{E}\hat{I}\sin(\theta_e - \theta_i)$
if $\phi = \theta_e - \theta_i$, then for lagging power factor ϕ is positive, and
$P + jQ = E\hat{I} = \bar{E}\hat{I}\cos\phi + j\bar{E}\hat{I}\sin\phi$

and for leading power factor ϕ is negative and the imaginary component is negative.

The Circle Diagram (Short Lines)

From the above, the power (per phase) at either end of a line is the product of line current and the conjugate of the voltage at the particular end. Let I_S be positive for the current flowing into the line, then the positive sending end power indicates power delivered to the line; and I_R be positive for current flowing out of the line, then positive receiving end power indicates power flowing out of the line.

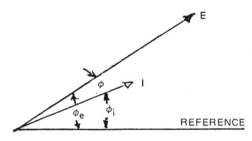

Figure 4-8. Diagram for Determining the Vector Equation for Power

and
$$I_S = I_R = I$$

$$P_S + jQ_S = \bar{E}_S I$$
$$P_R + jQ_R = \bar{E}_R I$$

Expressing current in terms of the terminal voltage:

$$I = \frac{E_S - E_R}{Z}$$

then
$$P_S + jQ_S = \frac{\hat{E}_S E_S - \hat{E}_R E_S}{Z}$$

$$P_R + jQ_R = \frac{-\hat{E}_R E_S + \hat{E}_S E_R}{Z}$$

In polar coordinate terms, the vectors become:

$$E_S = \bar{E}_S e^{j\theta_S} \qquad\qquad \hat{E}_S = E_S e^{-j\theta_S}$$

$$E_R = E_R e^{j\theta_S} \qquad\qquad E_R = \bar{E}_R e^{-j\theta_S}$$

$$Z = R + jX = Z e^{jY} \text{ where } \tan Y = \frac{X}{R}$$

and

$$\frac{1}{Z} = \frac{1}{Z e^{jY}} = \frac{e^{-jY}}{Z}$$

$$P_S + jQ_S = \frac{E_S^2}{Z} e^{-jY} - \frac{E_S E_R}{Z} e^{-jY} e^{-j(\theta_S - \theta_R)}$$

$$P_R + jQ_R = \frac{E_R^2}{Z} e^{-jY} + \frac{E_S E_R}{Z} e^{-jY} e^{j(\theta_S - \theta_R)}$$

Since only $(\theta_S - \theta_R)$ appear in the equation, let $\theta = \theta_S = \theta_R$ and expressing separately the real and imaginary parts of the power equation:

$$P_S = \frac{\bar{E}_S^2}{Z} \cos Y - \frac{\bar{E}_S \bar{E}_R}{Z} \cos (Y + \theta)$$

$$= \frac{\bar{E}_S^2 R}{Z^2} - \frac{\bar{E}_S^2 R}{Z^2} (R\cos\theta - X\sin\theta)$$

$$P_R = \frac{\bar{E}_R^{\,2}}{Z}\cos Y + \frac{E_S E_R}{Z}\cos(Y - \theta)$$

$$= \frac{\bar{E}_R^{\,2} R}{Z^2} - \frac{E_S E_R}{Z^2}(R\cos\theta + X\sin\theta)$$

$$Q_S = -\frac{\bar{E}_S^{\,2}}{Z}\sin Y + \frac{E_S E_R}{Z}\sin(Y + \theta)$$

$$= \frac{\bar{E}_S^{\,2} X}{Z^2} + \frac{E_S E_R}{Z^2}(R\sin\theta + X\cos\theta)$$

$$Q_R = -\frac{\bar{E}_R^{\,2}}{Z}\sin Y - \frac{E_S \bar{E}_R}{Z}\sin(Y - \theta)$$

$$= \frac{\bar{E}_R^{\,2} X}{Z^2} + \frac{E_S \bar{E}_R}{Z^2}(R\sin\theta + X\cos\theta)$$

The equation above for P_R indicates the maximum load that may be delivered at the receiving end will be maximum when $\cos(Y - \theta) = 1$, that is, when $Y = \theta$. The equation then becomes:

$$P_R(\text{max}) = -\frac{\bar{E}_R^{\,2}}{Z}\cos Y + \frac{\bar{E}_S \bar{E}_R}{Z} = \frac{E_R^{\,2}}{Z^2}R + \frac{\bar{E}_S \bar{E}_R}{Z}$$

In the equation for $P_S + jQ_S$ and $P_R + jQ_R$, when E_S and E_R are fixed in magnitude, the angle θ is the only variable. The first term of each of the equations is a fixed vector. The second term, added to the first, is fixed in magnitude but variable in phase. Plotted graphically, the expression $P + jQ$ (total) will thus describe a circle about the terminus of the fixed vector P as a center. These equations are shown graphically in Figure 4-9, where the real power is represented by the abscissa and reactive power by the ordinate of the coordinates.

The center of the sending end circle may be located by laying off the two components

$$\frac{\bar{E}_S^{\,2}}{Z^2}R \text{ and } j\frac{\bar{E}_S^{\,2}}{Z^2}X$$

in their proper direction. The end of the fixed vector determines the center of the sending end circle, which has the radius

$\dfrac{\bar{E}_S \bar{E}_R}{\bar{Z}}$. When $\theta = 0$, the vector $\left(\dfrac{\bar{E}_S \bar{E}_R}{\bar{Z}} e - jYe_e - j\theta \right)$ is parallel, but in opposite direction to $\dfrac{E_S^{\,2}}{\bar{Z}} e - jY$ (the fixed vector) the angle θ is measured from the fixed vector as shown in Figure 4 – 9.

The operating condition indicated by the given angle 0, the point A of the diagram shows the value of P_S and Q_S being delivered to the line at the sending end and the point B the value of P_R and Q_R delivered to the line at the receiving end. The difference between P_S and P_R is the I^2R loss of the line itself.

At each end, the value of Q is the reactive power that must be supplied to the line at the sending end or drawn from the line at the

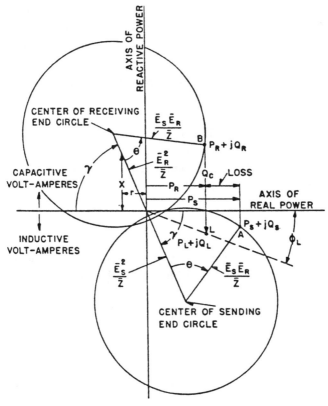

Figure 4-9. Power Circle Diagram for Short Lines. (*Courtesy Westinghouse Electric Co.*)

receiving end to maintain the desired terminal voltages. At the receiving end the reactive power drawn by the load at a particular power factor may not be equal to that needed to maintain the desired voltage. Some capacitance at the receiving end must be supplied to maintain the voltage. For instance, if the load P_L, indicated by point L in the figure, is to be supplied at the lagging power factor (shown by ϕ_L, then the inductive reactance volt-amperes indicated by Q_C must be supplied by capacitance.

It is to be noted that for a given circuit and desired voltage at both ends, there is a definite limit to the amount of power that may be transmitted; the critical value was shown above to be $\theta = Y$. The power limit may be increased (for the same current) by increasing the voltage at either or both ends. Increasing the voltage at one end increases the radius of both circles in direct proportion and moves the center only at that end away the origin along a line connecting the original center to the origin proportional to the square of the voltage at that end. Changes in the circuit will also change the power limit. A decrease in the magnitude of Z will result in an increase in the power that may be transmitted. Any change which decreases the series impedance will increase the power limit.

Circle Diagrams for Long Lines

The long line equivalent circuit may be represented by modifying the form of the short line equivalent circuit by the addition of the shunt capacitive reactance at each end.

$$Z'_{eq} = \overline{Z}'_{eq}e^{-jyu} = -jX'_{eq}$$

The equations for the terminal currents then have an additional term, as shown in the vector diagram of Figure 4-6 above.

$$I_S = \frac{E_S - E_R}{Z_{eq}} + \frac{E_S}{Z'_{eq}}$$

$$I_R = \frac{E_S - E_R}{Z_{eq}} + \frac{E_S}{Z_{eq}}$$

The equation for sending end power:

$$P_S + jQ_S = \left(\frac{\bar{E}_S^2}{\bar{Z}_{eq}} + \frac{\bar{E}_S^2}{\hat{Z}'_{eq}} \right) - \frac{\bar{E}_S\bar{E}_R e^{j\theta}}{\hat{Z}_{eq}}$$

and for the receiving end power:

$$P_R + jQ_R = \left(\frac{\bar{E}_R^2}{\bar{Z}_{eq}} - \frac{\bar{E}_R^2}{\hat{Z}'_{eq}} \right) + \frac{\bar{E}_R\bar{E}_S e^{j\theta}}{\hat{Z}_{eq}}$$

A comparison with similar equations for short lines shows them to be of the same form consisting of a fixed vector and a second vector, constant in magnitude but variable in phase, added to it. The power circle diagram can be plotted as shown in Figure 4-10.

In the above equation, the terms

$$\frac{\bar{E}_S^2}{\hat{Z}'_{eq}} \text{ and } -\frac{\bar{E}_R^2}{\hat{Z}'_{eq}}$$

are not a function of the angle θ and, hence, add directly to the "short line" fixed vector so that the effect is to shift the center of the power circles in the direction of volt-amperes only. The existence of the shunt reactances decreases the amount of positive reactive volt-amperes placed into the sending end of the line for a given amount of real power and increases the positive volt-amperes delivered at the receiving end. This decreases the amount of leading capacitive reactive volt-amperes that have to be supplied for a given load. It does not affect the real power conditions for a given operating angle or the load limit of the line. These factors are entirely determined by the series impedance of the line.

If the radius of the receiving end circle for $\theta = 0$ was plotted with the origin as the center, the vector would be at an angle Y with the real power axis. The angle shown in Figure 4-10 is therefore equal to Y, the angle of the equivalent series impedance. When $\theta - Y$ the maximum real power that can be delivered over the line.

The diagram for the sending end current is obtained from the power circle of the sending end and is referred to the vector of the sending end voltage. The diagram for the receiving end current is obtained from the power circle of the receiving end and is referred to the receiving end voltage.

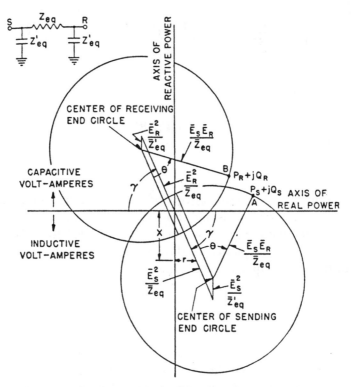

Figure 4-10. Power Circle Diagram for Long Lines
(Courtesy Westinghouse Electric Co.)

Loss Diagram

The resistance loss can be obtained from the power circle diagram, but more readily from the Loss Diagram.

$$\text{Loss} = P_S - P_R$$

Where the transmission line alone is under consideration:

$$
\begin{aligned}
\text{Loss} \;=\;& \frac{\overline{E}_S{}^2}{\overline{Z}^2}\,R - \frac{\overline{E}_S\overline{E}_R}{\overline{Z}^2}(R\cos\theta - X\sin\theta) \\[6pt]
&+ \frac{\overline{E}_R{}^2}{\overline{Z}^2}\,R - \frac{\overline{E}_S\overline{E}_R}{\overline{Z}^2}(R\cos\theta - X\sin\theta) \\[8pt]
=\;& \left(\overline{E}_S{}^2 + \overline{E}_R{}^2\right)\frac{R}{\overline{Z}^2} - 2\,\frac{\overline{E}_S\overline{E}_R}{\overline{Z}^2}R\cos\theta
\end{aligned}
$$

Figure 4-11 shows this relationship graphically.
    For the general equivalent circuit:

$$\text{Loss} = \left(\bar{E}_S^2 + \bar{E}_R^2\right)\frac{R_{eq}}{\bar{Z}_{eq}^2} + \frac{\bar{E}_S^2}{\bar{Z}_{S'}^2}R_{S'}$$

$$+ \frac{\bar{E}_R^2}{\bar{Z}_S^2}R_{R'} - 2\frac{\bar{E}_S\bar{E}_R}{\bar{Z}_{eq}}R_{eq}\cos\theta$$

This is equivalent to the formula (Figure 4-11) for the loss on the transmission line alone except for the terms which represent the losses in the resistance component of the shunt impedances $Z_{S'}$ and $Z_{R'}$:

$$\frac{\bar{E}_R^2}{\bar{Z}_S^2}R_{R'} \qquad \text{and} \qquad \frac{\bar{E}_R^2}{\bar{Z}_S^2}R_{R'}$$

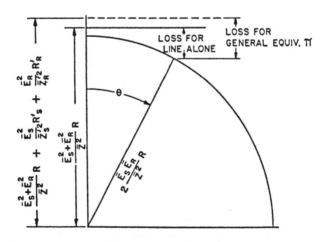

Fig. 4-11. The Transmission Line Loss Diagram
(when solving for general equivalent π loss,
substitute R_{eq} for R and $\bar{Z}_{eq}$ for $\bar{Z}$)
(Courtesy Westinghouse Electric Co.)

On the assumption of equal sending and receiving voltages, an equation for the load which can be delivered at a given permitted line loss. When loss is expressed as a percentage of P_R the equation is:

$$P_R = \frac{\% \text{ :pss}}{(100 + \% \text{ Loss})}\left(\frac{\overline{E}_R{}^2 \overline{X}_{eq}{}^2}{R_{eq}\overline{Z}_{eq}{}^2}\right)$$

and $\qquad Q_R = \overline{E}_R{}^2\left[\frac{X_{eq}}{\overline{Z}_{eq}{}^2}\left(1 + \frac{\% \text{ Loss}}{100}\right) - \frac{1}{X'_{eq}}\right]$

From the above equations, P_R is independent of the load power factor and the required amount of capacitance to maintain equal sending and receiving voltages for the delivered load P_R can be obtained by substituting the reactive kVA of the load from Q_R.

TRANSMISSION SUBSTATION
ARRANGEMENTS

The arrangement of transformers, circuit breakers and buses for transmission substations are generally similar to those for subtransmission and distribution substations, and for the same reasons of flexibility of operation and reliability of service. Some typical arrangements are shown in the one-line diagrams of Figures 4-1 and 4-12. The choice depending on the requirements of service continuity, the importance of which depends on the multiplicity of sources of supply and the type of load.

The busarrangement of Figure 4-13 is designed so that each incoming circuit supplies a fixed number of outgoing circuits, each independent of the other. Here a fault on one outgoing circuit does not interrupt service on the others, and likewise, a fault on one of the incoming circuits interrupts service on only the outgoing circuits it supplies. Nowhere are the incoming circuits connected in a grid. Here, interruption of supply from any one circuit will not communicate to another on both the incoming and outgoing sides; there will be no possible overloading of one circuit from attempting to pick up load supplied by the inter-

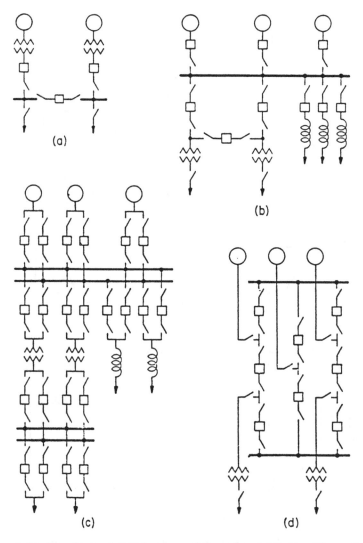

Figure 4-12. Fundamental Schemes of Supply at Higher Than Generated Voltage. *(Courtesy Westinghouse Electric Co.)*

rupted circuit, and hence, no cascading into total blackout. Obviously, there will be interruption of service to a given area, the load of which may or may not be picked up by other circuits at the command of the system operator, depending on the available spare capacity of the non-interrupted circuits that may be variable depending on the time of

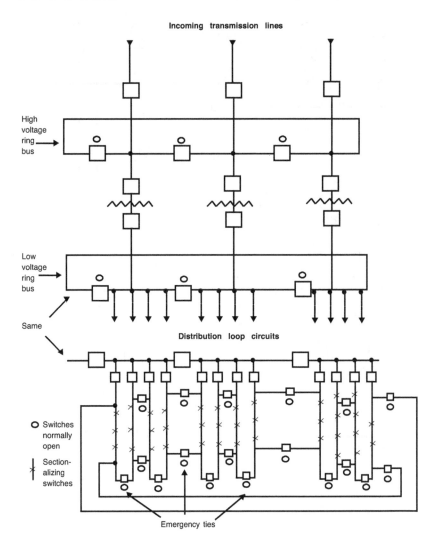

Figure 4-13. Substation Ring Bus Arrangement.

day. While reliability is thus affected, it limits the operations of the saboteur or vandal—the price in reliability paid for a higher degree of security.

Buses should be physically separated a sufficient distance so that failure of one with possible attendant explosion and fire, does not communicate to the others buses on other vital equipment. Similarly, circuit breakers and transformers should follow the same separation principle,

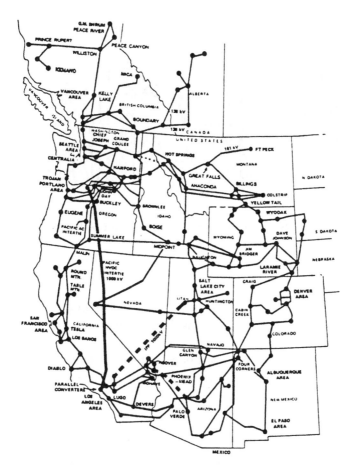

Figure 4-14. General Structure of the Western U.S. Power System.
(Courtesy Bonneville Power Administration)

achieved with steel reinforced explosion and fire proof barriers between them (Figure 4-15) (that may also act as sound barriers) and sump pits dug beneath each of these units sufficient to contain the oil in that unit even if aflame.

All equipment in the substation should be connected through air switches for safety reasons Figure 4-16). Whether for routine maintenance or emergencies, the worker must be able to see an opening in the circuit on both sides of the equipment on which he or she may be working.

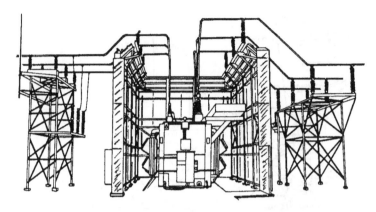

Figure 4-15. Substation Power Transformer with Barriers.

Like the other lines and equipment, the insulation requirements must take into account the basic insulation levels and coordination of insulation values.

Interconnection tie lines, part of a multi-company grid or pool, are treated as other sources or outgoing transmission lines.

POWER POOLS OR GRIDS

Transmission lines may provide interconnections for the transfer of power between two or more utilities for economic and emergency purposes. Sometimes referred to as integrated systems, they are commonly referred to as pools or grids, the interchange of power between utilities is to their mutual advantage. Example is shown in Figure 4-14.

Power interchange may take place not only between contiguous utilities, but even between utilities remotely situated from each other but part of the same pool or grid. Here, power is transmitted between the remote utilities using the facilities of other utilities between them. This is referred to as wheeling or wheel-barrowing of power. The intervening utilities are compensated for the use of their facilities with terms usually included as part of the contractual arrangements entered into when attaining membership in the pool or grid.

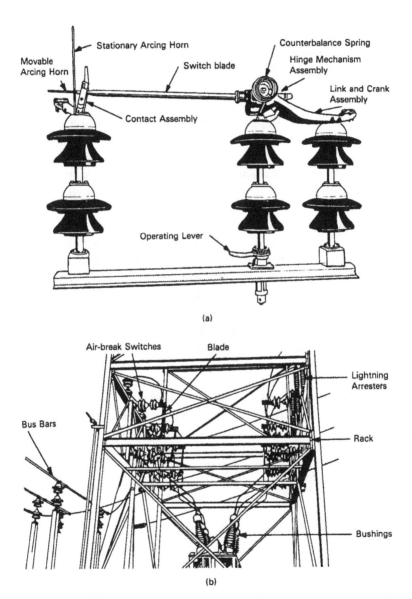

Figure 4.16. Air-break Switches Mounted on a Substation Rack.

NEW TRANSMISSION LINES

Prudence would suggest new transmission lines built during the period of national emergency not be tower lines as they are too obvious a target for saboteurs, not only from their exposure, but also because of their large capacity whose removal would have a marked effect on the available capacity in the chain of supply. Construction would be limited to low profile wood structures capable of rapid repair and replacement.

While temporary bypasses would be constructed to reenergize the high voltage tower lines, the conductors achieving the necessary code clearances horizontally, the acquisition and maintenance of such a wide swatch of right-of-way for any length would make it economically and environmentally preferable to repair and restore the tower line to its original condition; in some rare cases, it might be worth living with this temporary bypass until the national emergency is over.

The new low profile wood structures would impact on the voltage of the new lines and their capacity to meet Code clearances, the maximum height of the structure would be in the nature of some 70 feet, limiting voltage of the line to approximately 200 kV. This limitation also coincidentally applies to solid insulation cables that may in some instances be employed in place of or in extension of the main line. The 70 foot height also is in line with the range of bucket vehicle operation.

The capacity of existing transmission lines may be increased substantially by either raising the line voltage or by adding a second conductor to each phase of the circuit. These may be accomplished if the structures supporting the lines can retain code clearances by changing insulators or adding units to suspension strings; or the additional conductors producing additional ice and wind loadings do not overstress the conductor supports, Figure 4-19.

Figure 4-17. Insulated Bucket Vehicle.

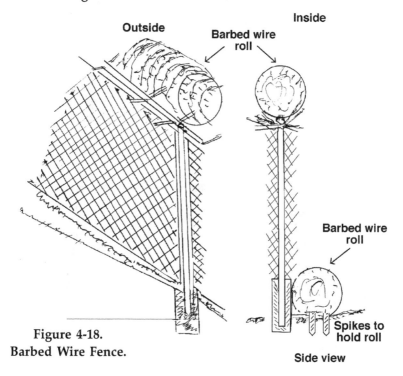

**Figure 4-18.
Barbed Wire Fence.**

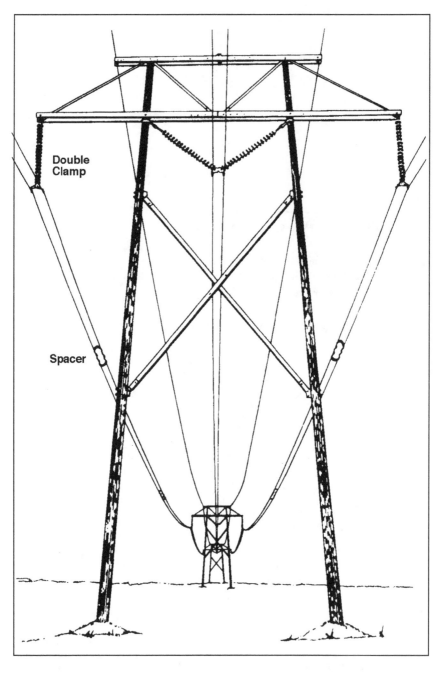

Double
Clamp

Spacer

Figure 4-19. Line installation in open, flat terrain.

Chapter 5

Electrical Protection

The protection of the several elements of a utility system is of paramount importance. Indeed, such protection is an integral part of the design of an electric system. In the design procedure, provision is made, as practical as possible, to prevent faults from happening, to limit the effects of a fault not only in avoiding or restricting damage it may cause, but also limiting the extent of the electric system that may be deenergized, and finally, to permit restoration of the affected elements as quickly as possible. This is generally accomplished by the installation of devices such as surge arresters, fuses, circuit breakers, corona and insulation guards, ground wires and grounds, and in particular instances, specification of the size and type of conductor to enable fault currents to burn it clear.

These applications provide electrical protection and are in addition to the mechanical protection provided by good construction and maintenance of lines and equipment, selection of good equipment, protective devices designed for particular installations, and such mundane items as locks and interlocks, fences and barriers, alarm systems, protective lighting, etc.

Protective relays play an important part in the operation of a transmission system. They initiate the opening of the breaker that may take only a fraction of a second to complete. The relay may take even less time to function, making the length of the circuit from the relay to the mechanism operating the breaker a factor to be considered. Where the coordination of operation of several breakers is involved, the relaying circuitry may be complex and prone to malfunctioning. In this case, redundant relaying should be considered. Where the breakers involved are distant from each other, communication between relays associated with them is necessary, and this may be accomplished by pilot wire, leased telephone wire, wireless, and, of recent usage, fiber optic conductors, may be considered. Electronic relays are much faster than the older electromagnetic type and retrofitting of such older relays is recommended.

In planning the circuitry of transmission systems, the associated relaying required to accomplish the desired results must always be taken into account. Sometimes, it may be the determining factor in the circuitry chosen.

The principle of the sectionalized open loop circuit is applicable to both transmission and distribution lines, the moveable open point allows transfer s of loads in both normal and emergency situations, limits service interruptions only to the section on which the fault occurs; the only time the loop is closed is momentary during switching times. The loop may originate and end at the same station, or may constitute an open tie between two stations. The opening in the circuit prevents disturbances (e.g. overloads, stability) from being communicated to other stations. Accomplishing this at stations by means of buses is detailed in Chapter 4, Transmission System Electrical Design.

THE DISTRIBUTION SYSTEM

Distribution Transformers

Starting with the distribution transformer on a radial primary feeder, the transformer is protected by a surge arrester and a fused cutout. The surge arrester protects the transformer and line from surges caused by lightning or switching. The fuse protects them from overloads (as well as surge currents) and fault currents as a result of faults on the secondary or in the transformer. The surge arrester and fuses are coordinated so that the arrester operates to drain the voltage surge before it can send sufficient current through the transformer. Both are coordinated with the insulation of the line and transformer, but principally with the latter.

In the special case of the so-called completely self-protected (CSP) transformer, the surge arrester performs the same function as above. The "weak link" within the transformer tank protects the unit from surge and fault currents, but not necessarily from overloads; circuit breakers on the secondary side are provided for this purpose, their opening coordinated with the weak link and surge arrester.

When distribution transformers are banked and their secondaries connected together through fuses, the fuses are meant to take care of overloads, although obviously the fuses would blow should fault current flow through them. The bank fuses are coordinated to blow before the fuses in the primary fused cutout meant to protect the transformer.

In the case of low voltage networks, there are fuses on the secondary mains known as limiters, and these are not meant to operate during overloads but to blow on fault current to limit the extent and severity of faults on the secondary mains designed to burn themselves clear. Fuses on the secondary side of the network transformer are there principally to blow when the protector fails to open when its primary feeder is deenergized; that is, to clear large currents not necessarily caused by faults. Their ratings, however, are not in the nature of other fuses designed for clearing faults. Where fuses exist on the primary side of the transformer, they serve to protect the transformer from the primary feeder should a fault occur in the transformer. All of these fuses are coordinated with the insulation values of the mains and equipment with which they are associated.

Fuses designed to blow on fault currents, but not on overloads, are sometimes referred to as current limiting fuses.

Primary Radial Feeders

Protection on a primary feeder includes fuses on laterals, and fuses, reclosers and circuit breakers on the main trunk and main branches. Fuses on the laterals must coordinate with those associated with distribution transformers and the devices associated with the primary main. The lateral fuse must blow after those of the distribution transformers and before the protective devices on the primary main operate. Refer to Figure 5-1.

On the main, the fuses, reclosers and circuit breakers must coordinate with each other. Those farthest from the source operate first and those next farthest operate next, and so on, to the circuit breaker at the substation. All of these operate on overloads, but obviously will also operate on fault current; they also coordinate with insulation values of the line and equipment. The recloser and circuit breaker are activated by overcurrent relays that usually include accessories that permit them to reclose automatically after an opening operation. Should the fault or overload be of a temporary nature (such as a limb or squirrel making temporary contact to ground or other energized conductors), the reclosing will reenergize the faulted circuit; typical settings include a first reclosure immediately, perhaps a cycle or less, that is, no intentional time delay; a second operation after a time interval of perhaps one or two seconds; and a third and final "lock out" of the recloser. Reclosers are actually circuit breakers of lower interrupting duty than those found in substations.

Often, when fuses are placed in the main trunk or branch to sectionalize a primary feeder, a "three-shot" fuse is used. Three fuses are mounted together, but only one is connected to the line. When it blows, a mechanism (including a time delay, if desired) operates to connect the second fuse to the line. Should this fuse blow, the operation is repeated, connecting the third fuse to the line. Should this fuse also blow, the line remains deenergized.

Throw-Over Supply

When two (or more) feeders are employed in a service to a consumer, a throw-over switch enables the transferring of the load to a second supply feeder should the first (or normal) supply feeder become deenergized for any reason; switching may include even a third supply feeder, if desirable. This may be done manually or automatically.

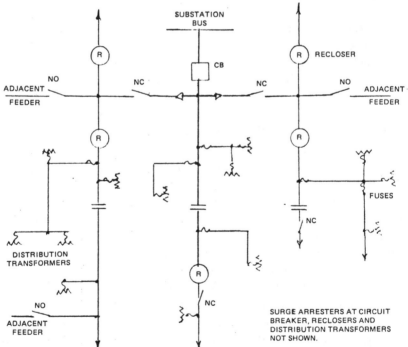

Figure 5-1. Radial Primary Feeder with Protective and Sectionalized Devices

Loop Primary Feeder

Open loop primary feeders are essentially radial feeders with a switching arrangement, usually a circuit breaker, at each end of the loop installed at the source substation or substations, or a circuit breaker on the line between the two ends of the loop. The breakers usually are operated after the fault is isolated by sectionalizing the branches of the loop employing disconnects normally manually operated.

The closed loop may employ additional circuit breakers to sectionalize the loop to isolate the fault. These breakers may be operated manually or automatically or automatically by relays that operate through pilot wires, to identify the direction of the fault current in the loop, the change in direction in adjacent circuit breakers indicating the location of the fault.

Primary Network

Here, the primary feeders forming the network are protected by circuit breakers installed at the several substations constituting the network.

This system of protection coordination may also be employed on the entire electric system back to the generating station, as shown in Figures 5-2 and 5-3. Protection schemes for substations and transmission lines, however, are somewhat different, and are treated separately.

Distribution Substation

Protective devices and equipment at the substation differ somewhat from those employed on the associated distribution feeders. For

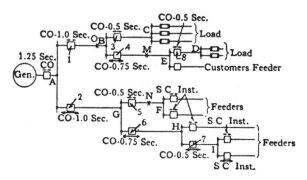

Figure 5 -2. Coordination of Overcurrent Protection on a Radial Power System. *(Courtesy Westinghouse Electric Co.)*

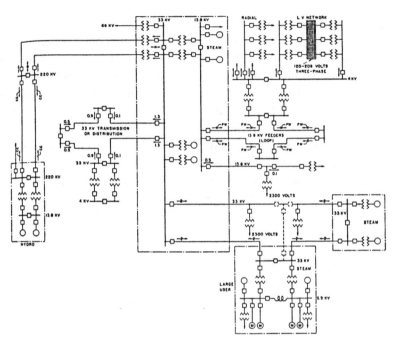

Figure 5-3. Composite Power System Illustrating Typical Protective Problems and Their Solution. *(Courtesy Westinghouse Electric Co.)*

the outgoing distribution feeders, circuit breakers are installed that operate to open under overload and fault conditions. The associated relays, in general, provide "overcurrent" protection, whose settings must coordinate with the other protective devices on the associated primary feeder.

The protection of the buses, the transformer, the circuit breakers on the incoming subtransmission or transmission feeders involve more complex types of relaying. These will be discussed before describing the protection to the several elements in the substation.

The protective relay receives data from the line or equipment it is protecting in small manageable quantities directly proportional to the actual quantities involved. These are generally values of current and voltage transmitted through current and potential instrument transformers. The relays receive continuous information of the conditions prevailing in the line or circuit with which they are associated. When abnormal conditions are sensed, the relays operate closing (or opening) contacts which complete (or interrupt) a circuit that, in turn, actuates machinery that operates (open or close) circuit breakers or other apparatus. They are low-powered devices used to activate high-powered ones. Practically all of the relays depend on the magnitude and direction of the currents and voltages involved. Mechanically, the relay contacts that are made or broken may be accomplished by plunger type, inductance type, or electronic type, described below.

The most common and simplest of the protective relays is the overcurrent relay, in which, at a predetermined value of current flowing in the line or equipment, the contacts close to activate the devices to operate the associated circuit breaker or other equipment.

The basic overcurrent relay simply operates to close or open its contacts as quickly as possible, that is, instantly, with no intentional delay, but experiencing some delay because of the time it takes for the device itself to operate mechanically, curve a in Figure 5-4. This may take from one-half to twenty cycles, and may result in actual settings higher than desirable to prevent frequent relay operations caused by

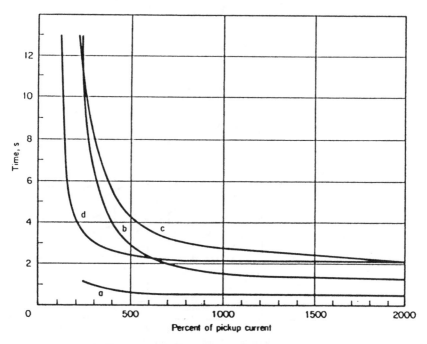

Figure 5-4. A Collection of Time Curves. These are representative of the various types of time curves which are used on overcurrent relays. (Courtesy Westinghouse Electric Co.)

transient nonpersistent conditions, Figure 5-5.

By modifying the elements of the instantaneous relay, including the restraint on the movable element, the time-current characteristic of the relay may be changed so that the greater the current, the shorter the time of operation of the relay. This is known as the inverse-time overcurrent relay, Figure 5-4 curves b and c, and provides greater flexibility in coordinating with fuses that may be in series with the breakers associated with the relay, Figure 5-6.

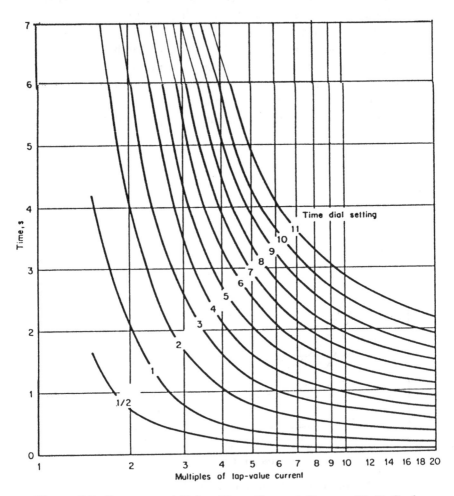

Figure 5-5. Overcurrent Relay Time-Current Curves, 50-60 Cycles
(Courtesy Westinghouse Electric Co.)

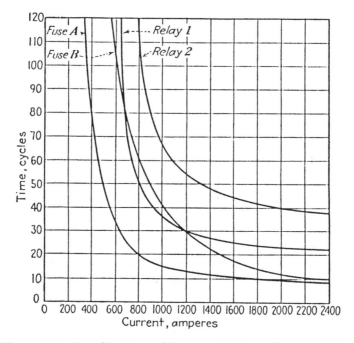

Figure 5-6. Coordination of Fuse and Relay Characteristics
(Courtesy Westinghouse Electric Co.)

To obtain greater selectivity, a definite time delay is introduced, Figure 5-4 curve d. This prevents abnormal currents of any value from operating the relay until after a definite time has elapsed. This is often combined with the inverse-time characteristic to obtain an inverse definite minimum time overcurrent relay. The flat part of this characteristic results in only a small increase in relay time for smaller fault currents, but simplifies greatly the coordination of relays.

The construction of the relays plays an important part in the operation. These may be classified as electromechanical or plunger induction or disc type, and electronic type. A brief description may prove useful in understanding their operation.

As the name implies, the plunger type consists of a steel plunger within a coil or solenoid, Figure 5-7. The current in the coil pulls up the plunger that causes contacts to be made or unmade: the greater the current the greater the speed of the plunger movement. The relay may be set to raise the plunger at any predetermined value by changing the position of the plunger and by taps on the coil. Inverse time delay is obtained by an oil-filled dashpot whose piston is attached to the plunger, the delay being governed by the size of the opening that permits oil to flow from one side to the other of the piston. Although the accuracy of this type is adequate, it is not as good as the later developed induction type. It is obsolete, but many exist.

The induction type is essentially a simple induction motor whose rotor is a metallic disc that rotates to close or open contacts, Figure 5-8. The torque that tends to turn the disc depends on the current flowing in the static coils; the tendency to turn is balanced against a spring which keeps it from turning until the current meets or exceeds predetermined limits. This type relay has an instantaneous characteristic. A reactor placed in the circuit of the relay introduces a time delay, giving the relay an inverse definite-minimum time characteristic. Settings may be changed by varying the distance the disc travels to close or open contacts, by changing the tension of the associated spring, and by taps on the relay coil. This type relay constitutes the greatest number in operation.

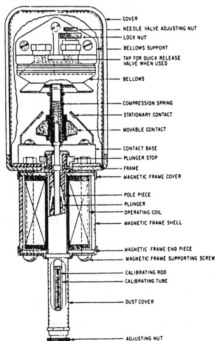

COVER
NEEDLE VALVE ADJUSTING NUT
LOCK NUT
BELLOWS SUPPORT
TAP FOR QUICK RELEASE VALVE WHEN USED
BELLOWS
COMPRESSION SPRING
STATIONARY CONTACT
MOVABLE CONTACT
CONTACT BASE
PLUNGER STOP
FRAME
MAGNETIC FRAME COVER
POLE PIECE
PLUNGER
OPERATING COIL
MAGNETIC FRAME SHELL
MAGNETIC FRAME END PIECE
MAGNETIC FRAME SUPPORTING SCREW
CALIBRATING ROD
CALIBRATING TUBE
DUST COVER
ADJUSTING NUT

Figure 5-7. Overcurrent Relay - Plunger Type
(Courtesy General Electric Co.)

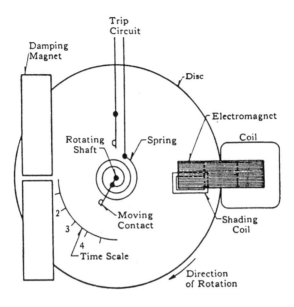

Figure 5-8. An Elementary Induction Type Relay
(Courtesy Westinghouse Electric Co.)

Electronic type relays use the solid-state control element as a switch that operates to energize circuits that activate the circuit breaker actuating devices. Relay contacts are not necessary which enables a more rapid response that, together with the greater accuracy obtainable, results in greater flexibility and selectivity in the protection systems. Time delay is achieved by controlling the charging and discharging of a capacitor through a resistor, resulting in the relay having the same characteristics as the induction type. The speed of operation of this type relay is greater than other types; moreover, it is possible for response of this relay to open the circuit breaker at a time when the fault current is at or near its minimum value in the cycle. (Both of these affect the short circuit duty of the circuit breaker.) The use of this type relay is expanding but economics prevents wholesale replacement of the inductive type relays in operation. Some basic electronic relay circuits are shown in Figure 5-9.

Returning to the protection of the elements making up the distribution substation, these may be classified into the protection of the transformers, the high-voltage bus (if any), the low-voltage bus, and the incoming and outgoing lines.

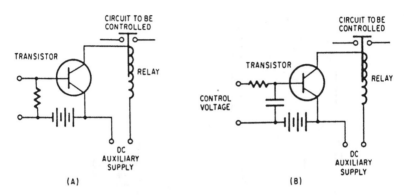

Figure 5-9. Basic Electronic Time-Delay Relay Circuit Using Resistance-Capacitance Combination. *(Courtesy General Electric Co.)*

Transformers

Transformer protection is accomplished in two ways: thermally and electrically. Thermal relays installed in the tank to measure the temperature of the conductors and of the surrounding oil give warning of impending failure. Through transducers, the temperature values are transmitted electrically to a control center where their supervision may include an auditory and visual alarm system. Electrically, the transformer is protected by a comparison of the input currents to the output currents: these two currents (taking into account the transformer ratio) should be equal (except for exciting current) in the transformer. These two currents are transmitted (through suitable current transformers) to two essentially overcurrent relays mounted together in one unit. Adjusted for the synchronizing current, these two relays buck each other so that no operation of the relay takes place under normal conditions. A failure within the transformer unbalances the currents transmitted to the relay causing it to operate to open the circuit breakers on both sides of the transformer, isolating it electrically. The relay action is called differential relaying, Figure 5-10, and the relay is called a differential relay.

Bus Protection

Bus protection is difficult to attain by relay action. Such a bus in a distribution substation has one or more incoming high voltage supply feeders and a number of (perhaps as many as eight) outgoing primary distribution feeders. On the high voltage side, circuit breakers may exist on one or both sides of the supply transformer; on the low voltage side,

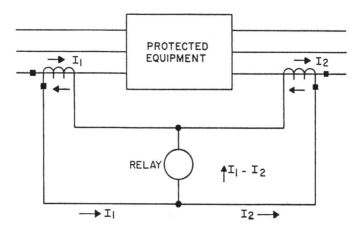

Figure 5-10. The Basic Differential Connection
(Courtesy Westinghouse Electric Co.)

each of the primary distribution feeders may have a circuit breaker associated with it.

To protect such a bus by differential relaying, involving the paralleling of current transformer secondaries on both sides of the bus is not very practical if a large number of feeders is involved; not only is it uneconomical, but the distortions that result may affect the accuracy of operation of the relay. Other schemes involving the measurement of bus impedances, or the direct current component of the fault current are, in general, also impractical.

Where the bus may be divided into two or more sections, with or without a circuit breaker tie between sections, differential relaying becomes practical. In any event, a fault on a bus or section of the bus, an interruption to a number of distribution feeders will occur. The installation of a "trouble" bus to which the distribution feeders may be automatically switched when a fault occurs on the "main" bus provides a better solution to the problem of bus protection. The relaying involved would include a simple throwover arrangement. (See Figure 2-50.)

Primary Distribution Feeder
    In almost all instances, primary distribution feeders are connected to the supply bus through circuit breakers. A fault on a feeder actuates an overcurrent relay that acts to open the associated circuit breaker.

Surge Arrester

Outdoor feeder exits, buses between transformers, circuit breakers, are particularly susceptible to voltage surges from lightning or switching, and arresters installed at strategic points are imperative. The characteristics of the arresters must be coordinated with those of the lines and equipment, both on the subtransmission and distribution sides, including the basic insulation levels.

Connections to ground from the arresters, as well as those from metallic structures, for safety reasons are extremely important. Not only must the electrical connections be mechanically continuous and sound, but the ground to which they are connected must be of sufficiently low resistance so that the surge voltage and associated energy be quickly and safely dissipated. Often, this may include a mesh of conductors buried one or more feet below the surface of the area in which the equipment is located, together with a multiplicity of ground rods interconnected with the mesh.

SUBTRANSMISSION CIRCUITS, SUBSTATION

Subtransmission circuits may be radial, parallel or loop circuits, or so interconnected to form a grid. These are similar to distribution type circuits and, for this discussion, their protection may be referred to that for distribution circuits.

The same observation may be made for protection of the elements at the subtransmission substation.

Some of their characteristics, however, are similar to those of transmission circuits and substations, and are discussed below.

TRANSMISSION CIRCUITS

The protection of transmission lines presents some special problems. The lines assumed are very long, perhaps 50 to over 100 miles, operating at voltages of 69 kV and above.

In shorter circuits, protection may be achieved by overcurrent relays at each end of the lines. A fault on a line will operate the directional element at one end and the overcurrent element at the other end of the line, no matter the location of the fault, to deenergize the line.

On a long transmission line, generally radiating from a generating source, the use of timed overcurrent relays (see Figure 5-4) as the means of obtaining selectivity results in the undesirable feature that the relays closest to the generating source have the largest time setting, although the clearing time should be as low as possible. To overcome this condition, the impedance or distance relay was developed. Essentially, the impedance element is a voltage restrained overcurrent relay which can be adjusted over a wide range to provide various time-distance characteristics. The time of operation of the relay is determined by the magnitude of the current and voltage applied to the relay. The higher the voltage, the greater the distance the current element must travel to overcome the restraining torque of the voltage element. In this way, the relay operates very fast for a close in fault since the voltage in the relay is nearly zero. Conversely, for a remote fault, the voltage will be higher and the current usually lower, so that a larger time is required for the relay contacts to close.

On long lines, a fault near one end of the line will operate the relay close to the fault first but, because of the relatively large impedance of the long line, there may not be sufficient fault current flow from the far end to operate the relay quickly to open the circuit breaker to disconnect the faulted feeder from the source. As mentioned, this is because the relay time setting at the relay of the transmission line closest to the generating source is made deliberately high to insure the proper operation of the other relays in succession back to the distribution substation. Often, the use of impedance relays is found to be unsatisfactory.

The ideal relay protection for transmission lines is the opening simultaneously at both ends of a line experiencing a fault. Distance relays, as outlined above, do not meet this criterion. One solution is to provide differential protection to the lines. This may be accomplished by the use of pilot wires or other communication channels.

The differential relaying can be accomplished via pilot wire channel as well as by carrier and microwave pilot channels. Protection zones, utilizing differential relaying, from the generating source to distribution substations, are shown in Figure 5-11; of particular interest are the transmission lines.

Pilot Wire Protection

Several methods of employing pilot wires are shown in Figure 5-12. The several schemes employ from two to six wires. Schemes employing current transformers at each end are shown in Figure 5-12, a, b, c, d

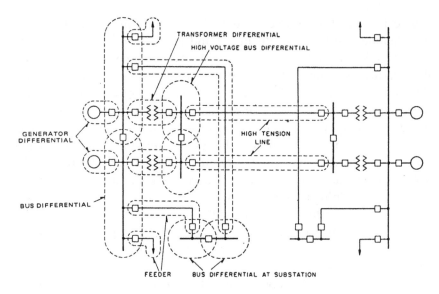

Figure 5 -11. Typical System Showing Protective Zones -Generation and Transmission. *(Courtesy Westinghouse Electric Co.)*

and f. A scheme employing differential relays atone end is shown in Figure 5-12 d, e. One scheme employing only two pilot wires, polyphase directional relays at each end, and a direct current source, is shown in Figure 5-12e. Another scheme, shown in Figure 5-12f, a simplified circuit that requires only two pilot wires, an alternating current source and special type relays that combine the currents in each of the current transformers into a single-phase voltage, is compared to a similar quantity from the opposite end of the line. Disadvantages of pilot wire include the over-burdening of current transformers, costs of leasing or installation of such systems, and more importantly, the practical limit of effective and positive operation of only some ten miles. For longer lines, carrier pilot relaying and microwave relaying schemes are employed.

Carrier Pilot Protection

In place of pilot wires, the inputs to the differential relays may be transmitted by a high-frequency current (50 to 200 kilocycles per second) superimposed on the conductors of the transmission line itself. The carrier signal normally operates the relays to keep the circuit breakers closed; if a fault occurs on the transmission line, the carrier signal is interrupted and the system "fails safe" with the opening of the circuit

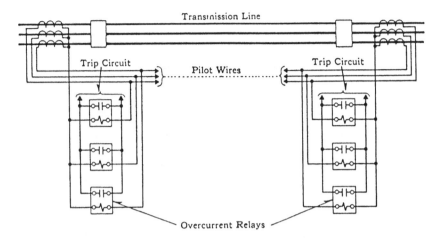

Figure 5-12a. Circulating Current Pilot Wire Scheme Load currents and through fault currents circulate over the pilot wires. *(Courtesy Westinghouse Electric Co.)*

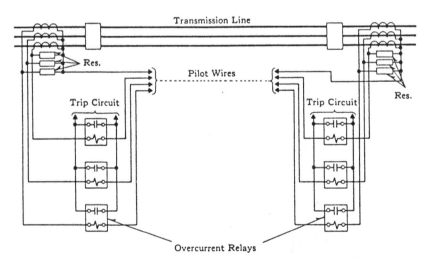

Figure 5-12b. Balanced Voltage Pilot Wire Scheme Load currents and through fault currents produce equal opposing voltages of the line terminals to prevent current flow in the relays and pilot circuit. *(Courtesy Westinghouse Electric Co.)*

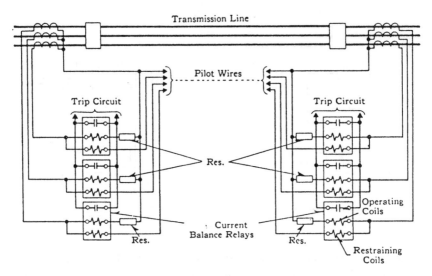

Figure 5-12c. Circulating Current Pilot Wire Scheme Using Current Balance Relays Secondary currents must be kept low to keep the burden low. *(Courtesy Westinghouse Electric Co.)*

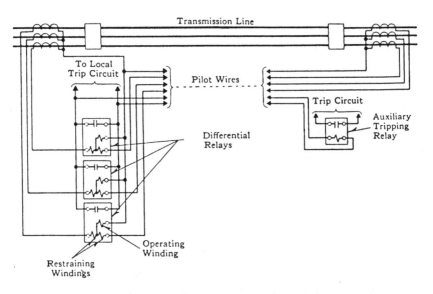

Figure 5-12d. Pilot wire Scheme Using Percentage Differential Relays. Note the similarity in the connections compared to apparatus protection using differential relays. *(Courtesy Westinghouse Electric Co.)*

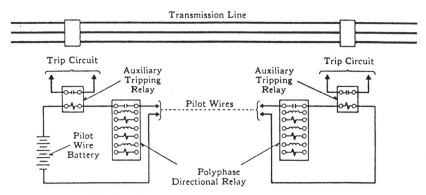

Figure 5-12e. The Directional Comparison Pilot Wire Scheme Direct current is used over a pair of wires. The alternating current connections are omitted for simplicity. *(Courtesy Westinghouse Electric Co.)*

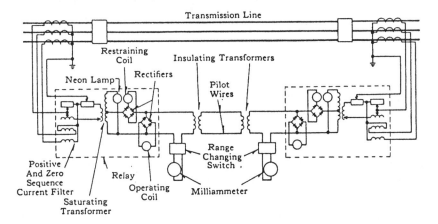

Figure 5-12f. Alternating Current Pilot Wire Scheme Using Sequence Relays Only two pilot wires are needed. *(Courtesy Westinghouse Electric Co.)*

breakers. A simplified diagram is shown in Figure 5-13. Carrier systems operate effectively and positively over several hundred miles. Moreover, the channel can be used for other purposes: telemetry, supervisory controls, telephone communication, and other related purposes.

Microwave Pilot Protection

In this wireless system, the pilot protection includes transmission of the associated signals over microwave radio channels. Such systems

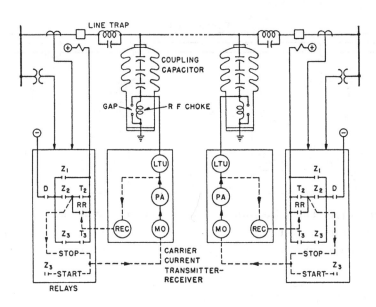

Dotted lines indicate symbolically the carrier controls
MO—Master Oscillator   REC—Receiver
PA—Power Amplifier   LTY—Line Tuning Unit

Figure 5-13. Carrier Current Relay System Including Relays, Carrier Current Transmitter-Receivers, Coupling Capacitors, and Chokes. (*Courtesy Westinghouse Electric Co.*)

are not affected by faults in the transmission line, and are capable of accommodating other separate functions. However, microwave radio is limited to line of sight transmission and changes in direction of the transmission line requires intermediate units to receive and retransmit the signals to the next unit, sometimes referred to as microwave relay stations.

Generally, all of the pilot schemes are designed to fail safe as described above.

Ground Relay

One other simple, protective relay is particularly adapted to three-phase systems. The currents flowing in each of the three energized conductors generally are fairly well balanced in magnitude so that there is little or no current in the return or ground, or neutral, conductor. Measuring the current in each energized conductor and determining the resultant (vectorially) difference, or measuring the ground or return current directly, this ground current can be made to operate a relay when it exceeds a predetermined value. A ground relay detecting faults is shown in Figure 5-14. This protection scheme is able to discriminate between load current and fault current.

Other protective schemes generally employ one or more of the systems described above. While the relays described above are based on induction type relays, electronically operated relays, some with no moving parts, accomplish the same purposes.

FAULT CURRENT CALCULATION

Faults on three-phase transmission lines often do not occur on all three phases simultaneously. Many times, the fault occurs on one phase to ground, or between two of the phases, between two phases and ground, and even the three-phase faults may or may not be to ground. Solutions and calculations may be determined by the method of symmetrical components, detailed in Appendix B.

STABILITY

When transmission lines are connected in a pool or grid served from two or more sources, loads may be apportioned among them according to a schedule that may take into account economics as well as reserves for contingencies (service reliability). When a fault on one of the

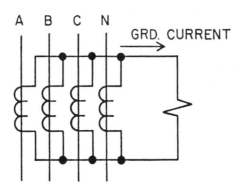

Figure 5-14. Ground Relay. *(Courtesy Westinghouse Electric Co.)*

lines occurs, the normal flow of current in each of the feeders will be disturbed and the current flow to the fault (fault current) will be proportioned among all the feeders depending on the impedance of the circuits between the sources of supply (generators) and the fault.

The generators closest to the fault electrically will attempt to pick up the greater part of the fault current over the lines to the fault and affect the flow of current in the other feeders connected in the grid. As these currents (load and fault) are imposed on the several generators, it will cause the generators then to slow down, but not an equal amount, the one supplying the greatest share of the current will slow down the most, while others may slow down proportionately and hence may no longer remain in step. The generator that slows down the least will attempt to pick up the greatest share of the load and fault current which will cause it to slow down, while the others, thus relieved, may tend to regain speed and again pick up more of the current flow. The net effect is for these generators to slow down and then speed up, creating a rocking motion. If the governors on the generators do not respond quickly enough, this rocking motion will tend to become aggravated. Unless the rocking motion is dampened, one of the generators may speed up (in relation to the others) to attempt to pick up so much more of the current flow that its protective overcurrent relays operate to trip one or more of the feeders it supplies.

The loss of one transmission line will cause the others to pick up the load and fault current it has dropped, and may cause other lines to become overloaded and trip from relay action. The cascading that results will eventually cause all of the transmission lines in the grid to open and the grid to shut down into a blackout. Settings on the protective relays and devices are so designed to permit additional loads to be carried by the feeders for planned contingencies (sometimes to account for two feeders out of service at the same time) but to segregate the faulted feeder as rapidly as possible, causing the affected generators to return to normal.

Chapter 6

Direct Current Transmission

Although there are relatively few direct current transmission lines in operation mainly because of the high cost of transforming alternating current into direct current at one end of the line and then back again to alternating current at the other end, and in a process that allows this to happen when current is flowing in either direction. However, it has so many other advantages, compared to alternating current, that it becomes affordable in certain instances.

One of the chief advantages is that one conductor may take the place of three in a-c conductors in a circuit. The return path for such a d-c circuit can be the ground, aided by the grounds associated with the lightning protection of the line, namely, the overhead ground wire and the underground counterpoises, some of which may involve a continuous path between structures (Figure 6-1). Crossing bodies of water, particularly sea water, the sea provides the return conductor. Other important advantages stem from the fact that d-c circuits do not have alternating magnetic and electrostatic fields about them, hence no problems with inductive and capacitive reactances with their effects on voltage and power losses that may seriously reduce the active power transmission capability, requiring the use of corrective reactors. For the same voltage rating as a comparative a-c circuit, the d-c circuit requires only some 70 percent of the insulation (Figure 6-2) or, expressed differently, the same insulation as the a-c circuit can accommodate a voltage some 30 percent higher with an increase of the capability of the d-c circuit to deliver some 30 percent more power. In connecting two d-c circuits of the same voltage rating, there is no need for synchronization; in similar a-c situations, where the a-c circuits are of different frequencies (e.g. 50 and 60 cycle), conversion to d-c- makes their connection feasible. All of which make the control of d-c circuits simpler than its a-c counterpart.

Care, however, should be taken in energizing and deenergizing d-c circuits as the rise and collapse of its associated magnetic fields may induce unwanted voltages in the conductor itself and in surrounding conductors. An automatic placing of temporary grounds on the switches during this part of their operation is usually provided to insure the safety of the workers. Care should also be taken not to have the d-c-circuit in close proximity to an energized a-c circuit as the a-c voltages induced in the d-c circuit may have some influence on the d-c "wave" and the existence of the a-c voltage will produce circulating currents that will only serve to heat the conductor and reduce the circuit capability; a-c filters are available that drain these unwanted circulating currents to ground.

A simplified schematic diagram of a typical d-c transmission line is shown in Figure 6-3. The a-c to d-c voltage rectifiers are usually thyrite units that are capable of rectifying a-c to d-c and, as inverters, converting the d-c back to a-c. The losses involved in their operation are relatively low and their maintenance (by replacing of worn out units) is also relatively low. Figure 6-4 shows a typical wave-form that is an example of change from maximum positive a-c voltage to d-c voltages. Some idea of the size of the banks of thyrite rectifier-inverters for a 345 kV circuit is shown in Figure 6-5. Clearly, what is needed is a d-c transformer that approximates the simplicity and efficiency of the a-c transformer.

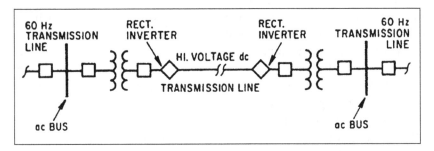

Figure 6-1. DC transmission used over very long distances. The two AC buses may be hundreds of miles apart and do not have to be synchronized, or in phase, to permit power to flow between systems.

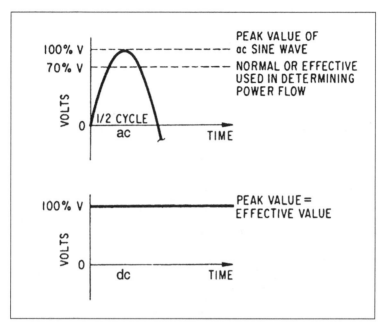

Figure 6-2. AC vs. DC power transmission. In AC, the peak voltage must be used in calculating the insulation required from conductor to grounded supporting structure. This value is higher than the effective value of the DC system shown at the bottom. The DC system utilizes the maximum voltage to ground to transmit power.

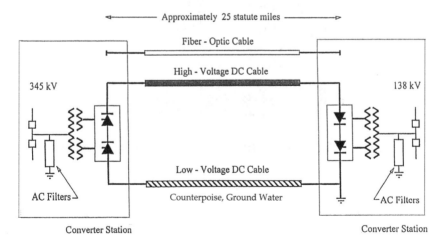

Figure 6-3. Simplified schematic diagram of a high voltage DC transmission line.

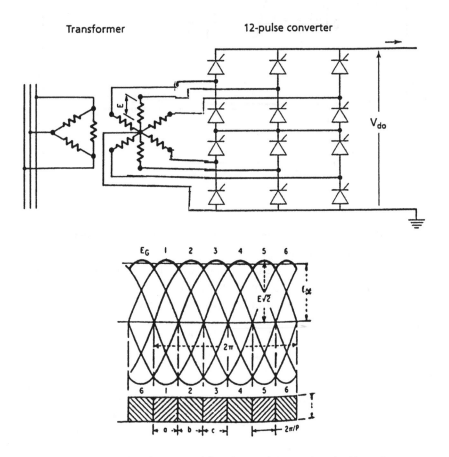

Figure 6-4. Six-phase rectification with twelve half cycles.

Figure 6-5.

Chapter 7

Overhead Mechanical Design and Construction

The design and construction of overhead lines, and their several parts, must be such that, in addition to normal stresses and strains, they sustain safely abnormal conditions caused by nature and people. Supports for conductors and equipment must withstand the forces imposed on them, and the conductors themselves must be strong enough to support the forces imposed on them, including their own weight.

National minimum standards for the design of overhead systems are established in the National Electric Safety Code (NESC) by the Institute of Electrical and Electronic Engineers (IEEE). These standards conform with those of other national bodies, including the American National Standards Institute (ANSI), the American Standards Association (ASA), the National Electrical Manufacturers Association (NEMA), and the American Society for Testing and Materials (ASTM), among others. The NESC standards have received acceptance by the utilities and other groups in the United States and elsewhere.

Generally, the NESC specifies:

1. Minimum separation or clearances not only between conductors, but also between conductors and surrounding structures for different operating voltages and under varying local load conditions.

2. Minimum strength of materials and safety factors used in the design and construction of proposed structures.

3. Loadings imposed by ice and wind on conductors and structures because of probable adverse climatic conditions, roughly defined by geographic areas.

The geographic areas designated by NESC are shown in Figure 7-1, dividing the country into light, medium and heavy districts; Hawaii and Alaska are assigned to the light and heavy districts respectively. The loading conditions are included in Table 7-1. The loading districts are only approximate and values for design purposes should be modified by other practical considerations such as local codes and regulations, environmental requirements, public relations, and other deviations based on experience.

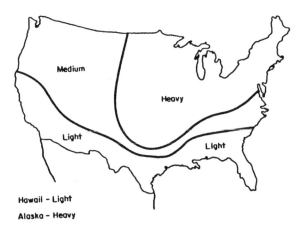

Figure 7-1. Loading Districts - NESC

STRUCTURES

The structures that support the conductors and equipment of an overhead system consist almost entirely of poles and towers. These structures are subject to vertical loading from the weight they must carry, that is, the conductors, crossarms, insulators, equipment and associated hardware, and ice that may form about them. But, more importantly, they are subject to horizontal forces applied near the top of the structure as a result of the pressure of the wind blowing against the ice covered conductors and equipment, from offsets in the line, and from uneven spans. Figure 7-2a, 7-2b, 7-2c.

The vertical loading represents dead weight of the items described above and exerts a compressive stress that may be considered uniformly distributed over the cross section of the pole or among the metallic

Table 7-1. Ice and Wind Loadings on Overhead Systems - NESC

Type of Loading	Radial Thickness of Ice		Wind Load on Projected Area of Conductors		Temperature	
	in.	*cm*	*lb/ft²*	*kg/m²*	*°F*	*°C*
Light	0.00	0.00	9	44	+30	- 1.1
Medium	0.25	0.63	4	20	+15	- 9.4
Heavy	0.50	1.27	4	20	0	-18.0

(steel) members of the tower. The horizontal loading requirements are generally so much greater that the vertical loading requirements are more than met by the mechanical requirements of the horizontal loading and usually are not given further attention.

Overhead distribution systems supporting structures consist almost entirely of poles (wood, metal, concrete) while transmission structures may be made of wood or steel, as are also substation structures.

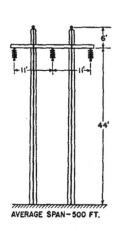

Figure 7-2a. Typical configurations of wood-pole lines.

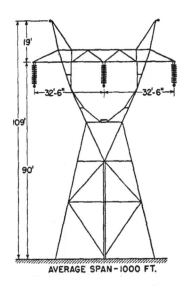

Figure 7-2b. Typical configurations of steel-tower lines.

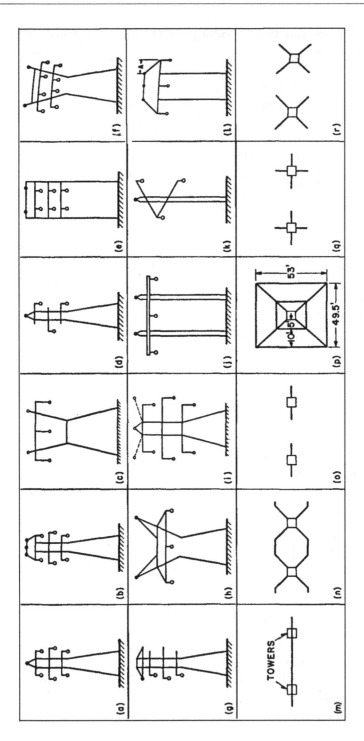

Figure 7-2c. Typical transmission structure and counterpoise configurations.

Poles

A pole is essentially a cantilever beam attached at one end and a load applied at the other end constituting the horizontal loading. The analysis for such a beam applies:

The bending moment causes stresses in the material (wood), tensile on the side opposite to that on which the load is pulling and compressive on the other side, shown in Figure 7-3.

Bending Moment (M) = Ph
Maximum Fiber Stress – at any cross section

$$f = M\frac{c}{I}$$

where   P   is   the applied horizontal force
              h   is   the perpendicular distance to the point where failure
                    may occur, usually the ground line

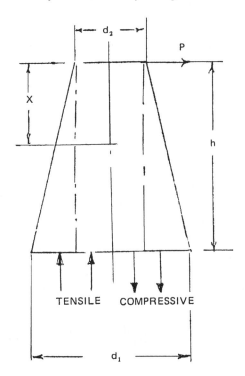

Figure 7-3. Pole Stresses and Configuration

c   is   the distance from the extreme fibers of cross section
         to the neutral axis

I   is   the moment of inertia of the cross section

d   is   the diameter for a circular cross section

then $c = 1/2\,d$

$$I = \frac{\pi d^4}{64} = 0.0491\,d^4$$

and $$f = \frac{M}{0.0982\,d^3}$$

and $$\frac{f}{c} = \frac{\pi d^3}{32} = 0.0982\,d^3 \text{(the section modulus)}$$

Wind Pressure

In determining the total bending moment on the pole, the pressure P on the length of the conductor with its coating of ice and its distance from the ground (at which the circular cross section is -to be determined) must be considered. The total moment is the sum of the moments of the several conductors plus the moment of the pole itself.

The projected area of the pole may be resolved into a rectangle and a triangle:

The pressure on the rectangle $= p_1 d_2 h_1$ where p_1 is the wind pressure in lbs/in^2.

The moment about the base $= p_1 d_2 \dfrac{h_1}{2}$

The pressure on the triangle $= p_1(d_1 - d_2)\dfrac{h_1}{2}$

its moment about the base $= p_1(d_1 - d_2)\dfrac{h_1^{\,2}}{6}$

Total pressure $= p_1 h_1 \dfrac{(d_1 + d_2)}{2}$

Total moment (due to wind) $= p_1 h_1^{\,2}\left(\dfrac{d_2}{2} + \dfrac{d_1}{6} - \dfrac{d_2}{6}\right)$

$$= p_1 h_1^{\,2}\left(\frac{d_2}{3} + \frac{d_1}{6}\right) \text{ inch pounds}$$

The weakest section of a pole of uniform taper and unit strength, theoretically, is at the point where the resultant load is applied, that is, the total moment divided by the total load above the ground. Referring to Figure 7-3

P   = total load applied where diameter is d_2

d = diameter of weak section at distance x from P

$$t = \text{taper of pole} = \frac{d_1 - d_2}{h_1} = \frac{d - d_2}{x}$$

Bending stress at d (the weak section)

$$f_d = \frac{P_x}{0.0982(d_2 + tx)^3}$$

For a maximum value of f_d the first derivative $\dfrac{df}{dx} = 0$

$$\frac{df}{dx} = d_2 + tx - 3tx = d_2 - 2tx = 0$$

$$d_2 = 2tx = 2(d - d_2)$$

$$d = \frac{3}{2}d_2$$

that is, the weakest cross section occurs at a point above the ground where the diameter is 1-1/2 times the diameter where the resultant load is applied. The stress at that point is:

$$f_d = \frac{P_x}{0.0982\, d^3} = \frac{P_x}{0.0982\left(\frac{3}{2}d_2\right)^3}$$

$$x = \frac{d - d_2}{d_1 - d_2}h_1 = \frac{d_2}{2(d_1 - d_2)}h_1$$

Substituting and expressing x in terms of h_1

$$f_d = \frac{P d_2 h_1}{0.0982\left(\frac{3}{2}d_2\right)^3 \cdot 2(d_1 - d_2)}$$

$$= \frac{P h_1}{0.662(d_1 - d_2)d_x^2}$$

The maximum unit stress at the weakest point is given in terms of diameter at the point of application of resultant load, diameter at base, and total moment at base.

The above is true only when the weak section is above the ground line. If the weak section is assumed at the ground line, the stress is:

$$f_d = \frac{Ph_1}{0.0982d_1{}^3} = \frac{M_t}{0.0982d^3}$$

where M_t is the total moment at the ground line, and is equal to the sum of the moments of the wind loads on all conductors and on the pole itself plus any other wind loads on equipment, etc., that may be present. For practical purposes, the assumption that the ground line is the weakest section is sufficiently accurate for design purposes, as poles are not exactly uniform in taper, cross section, and strength.

Where the loading is due to tension in conductors rather than wind loading, the same principles in obtaining moments apply, using tensions rather than wind pressures for loads due to the conductors.

To care for unknown or unforeseen conditions that may create stresses greater than the worst probable stresses determined, as described above, factors of safety are applied and are included in the NESC, Table 7-2. Various "grades of construction" are specified depending on field conditions that include the voltage of the lines, their proximity to other structures and communication lines, crossings of railroads, and main and secondary roads, urban or rural districts, etc. The NESC calls for varying safety factors, not only for poles at the time of initial installation, but at replacement. The latest revision of the NESC should be consulted before designs become final.

Table 7-2. Ultimate Bearing Strength of Wood, lb/in.^{2}

Wood	End-grain bearing	Cross-grain bearing
Long-leaf yellow pine	5000	1000
Douglas fir	4500	800
Western red cedar	3500	700
Cypress	3500	700
Redwood	3500	700
Northern white cedar	3000	700

Pole Stability

The stability of poles, in large part, depends on the depth of their setting. Certain minimum depths are essential if the poles are to develop their full strength. NESC and ASA recommendations are listed in Tables 7-3a and 7-3b. Deeper settings, however, should be used where poles may be under extra heavy stress, such as at corners, than for poles in a straight line.

Concrete Poles

Concrete poles have received wide acceptance, mainly because of appearance, including the installation of electrical risers within the hollow structure out of sight and not accessible to the public. With few solid exceptions, concrete poles are of hollow construction, in round, square, and polygon shapes. While the concrete is being poured, the forms are spun, forcing the concrete to the outside around the steel reinforcement, producing a highly uniform, compact, prestressed concrete of high strength and texture. Although heavier than wood, improved field methods have simplified their installation. Their stresses are determined in a similar manner as for wood poles, and they are set in the same manner. Their characteristics are shown in Tables 7-4a and 7-4b. They are used on both transmission and distribution lines, similar to wood pole lines.

Table 7-3a. Pole Setting Depths in Soil and Rock (NESC)

Length of Pole in Feet	Setting Depth in Soil in Feet	Setting Depth in Rock in Feet
20	5	3
25	5	3.5
30	5.5	3.5
35	6	4
40	6	4
45	6.5	4.5
50	7	4.5
55	7	5
60	7.5	5
65	8	6
70	8	6
75	8.5	6
80	9	6.5

Table 7-3b. ASA Standard Pole Dimensions and Depth Settings

Pole length, ft	Depth setting, ft	Wood*	Minimum circumference at 6 ft from butt (approximate ground line), in								
			Class and minimum top circumference, in								
			00 29	0 28	1 27	2 25	3 23	4 21	5 19	6 17	7 15
20	4	P	—	—	31.5	29.5	27.5	25.5	23.5	22.0	20.0
		C	—	—	34.5	32.0	30.0	28.0	25.5	23.5	22.0
		W	32.0	30.0	28.5	26.5	25.0	23.0	21.0	—	—
25	5	P	—	—	34.5	32.5	30.0	28.0	26.0	24.0	22.0
		C	—	—	38.0	35.5	33.0	30.5	28.5	26.0	24.5
		W	35.0	33.0	31.0	29.0	27.0	25.5	23.5	—	—
30	5½	P	—	—	37.5	35.0	32.5	30.0	28.0	26.0	24.0
		C	—	—	41.0	38.5	35.5	33.0	30.5	28.5	26.5
		W	38.0	36.0	33.5	31.5	29.5	27.5	25.5	—	—
35	6	P	45.0	42.5	40.0	37.5	35.0	32.0	30.0	27.5	25.5
		C	49.0	51.5	43.5	41.0	38.0	35.5	32.5	30.5	28.0
		W	40.5	38.0	36.0	33.5	31.5	29.0	27.0	—	—
40	6	P	47.0	44.5	42.0	39.5	37.0	34.0	31.5	29.0	27.0
		C	51.0	48.5	46.0	43.5	40.5	37.5	34.5	32.0	—
		W	42.5	40.5	38.0	35.5	33.5	31.0	28.5	—	—
45	6½	P	50.0	47.0	44.0	41.5	38.5	36.0	33.0	30.5	28.5
		C	54.0	52.0	48.5	45.5	42.5	39.5	36.5	—	—
		W	45.0	42.5	40.0	37.5	35.0	32.5	30.0	—	—
50	7	P	—	—	46.0	43.0	40.0	37.5	34.5	32.0	29.5
		C	—	—	50.5	47.5	44.5	41.0	38.0	—	—
		W	46.5	44.0	41.5	39.0	36.5	34.0	31.5	—	—
55	7½	P	—	—	47.5	44.5	41.5	39.0	36.0	33.5	—
		C	—	—	52.5	49.5	46.0	42.5	39.5	—	—
		W	48.5	46.0	43.0	40.5	38.0	35.0	32.5	—	—
60	8	P	—	—	49.5	46.0	43.0	40.0	37.0	34.5	—
		C	—	—	54.5	51.0	47.5	44.0	—	—	—
		W	50.0	47.5	44.5	42.0	39.0	36.5	33.5	—	—
65	8½	P	—	—	51.0	47.5	44.5	41.5	38.5	—	—
		C	—	—	56.0	52.5	49.0	45.5	—	—	—
70	9	P	—	—	52.5	49.0	46.0	42.5	39.5	—	—
		C	—	—	57.5	54.0	50.5	47.0	—	—	—
75	9½	P	—	—	54.0	50.5	47.0	44.0	—	—	—
		C	—	—	59.5	55.5	52.0	48.5	—	—	—
80	10	P	—	—	55.0	51.5	48.5	45.0	—	—	—
		C	—	—	61.0	57.0	53.5	49.5	—	—	—
85	10½	P	—	—	56.5	53.0	49.5	—	—	—	—
		C	—	—	62.5	58.5	54.5	—	—	—	—
90	11	P	—	—	57.5	54.0	50.5	—	—	—	—
		C	—	—	63.0	60.0	56.0	—	—	—	—

*P = yellow pine; C = western red cedar; W = wallaba.

Table 7-4a. Dimensions and Strengths—Round Hollow Concrete Poles

Overall length						*Design*		
m	*ft-in*	*Class*	*Setting depth, ft-in*	*Top diameter, in*	*Butt diameter, in*	*Design ultimate moment, ft · lb*	*Allowable moment—SF = 2, ft · lb*	*Nomina weight, lb*
12	39-4	A	5-11	$7\frac{1}{2}$	$13\frac{13}{16}$	100,530	50,260	2540
13	42-8	A	6-3	$7\frac{1}{2}$	$14\frac{5}{16}$	110,130	55,060	2830
14	45-11	A	6-7	$7\frac{1}{2}$	$14\frac{13}{16}$	119,470	59,730	3130
15	49-2	A	6-11	$7\frac{1}{2}$	$15\frac{3}{8}$	128,800	64,400	3470
16	52-6	A	7-3	$7\frac{1}{2}$	$15\frac{7}{8}$	138,400	69,200	3800
12	39-4	B	5-11	$7\frac{1}{2}$	$13\frac{13}{16}$	84,830	42,410	2480
13	42-8	B	6-3	$7\frac{1}{2}$	$14\frac{5}{16}$	92,930	46,460	2760
14	45-11	B	6-7	$7\frac{1}{2}$	$14\frac{13}{16}$	100,800	50,400	3060
15	49-2	B	6-11	$7\frac{1}{2}$	$15\frac{3}{8}$	108,680	54,340	3380
16	52-6	B	7-3	$7\frac{1}{2}$	$15\frac{7}{8}$	116,780	58,390	3710
9	29-6	C	4-11	$6\frac{11}{16}$	$11\frac{7}{16}$	50,810	25,400	1520
10	32-10	C	5-3	$6\frac{11}{16}$	$11\frac{15}{16}$	57,560	28,780	1770
11	36-1	C	5-7	$6\frac{11}{16}$	$12\frac{7}{16}$	64,130	32,060	2010
12	39-4	C	5-11	$7\frac{1}{2}$	$13\frac{13}{16}$	70,690	35,340	2430
13	42-8	C	6-3	$7\frac{1}{2}$	$14\frac{5}{16}$	77,440	38,720	2710
14	45-11	C	6-7	$7\frac{1}{2}$	$14\frac{13}{16}$	84,000	42,000	3000
15	49-2	C	6-11	$7\frac{1}{2}$	$15\frac{3}{8}$	90,560	45,280	3310
16	52-6	C	7-3	$7\frac{1}{2}$	$15\frac{7}{8}$	97,310	48,650	3640
9	29-6	D	4-11	$6\frac{11}{16}$	$11\frac{7}{16}$	41,780	20,890	1490
10	32-10	D	5-3	$6\frac{11}{16}$	$11\frac{15}{16}$	47,330	23,660	1720
11	36-1	D	5-7	$6\frac{11}{16}$	$12\frac{7}{16}$	52,730	26,360	1960
12	39-4	D	5-11	$7\frac{1}{2}$	$13\frac{13}{16}$	58,120	29,060	2390
13	42-8	D	6-3	$7\frac{1}{2}$	$14\frac{5}{16}$	63,670	31,840	2670
14	45-11	D	6-7	$7\frac{1}{2}$	$14\frac{13}{16}$	69,070	34,540	2960
15	49-2	D	6-11	$7\frac{1}{2}$	$15\frac{3}{8}$	74,460	37,230	3260
16	52-6	D	7-3	$7\frac{1}{2}$	$15\frac{7}{8}$	80,010	40,000	3570
9	29-6	E	4-11	$6\frac{11}{16}$	$11\frac{7}{16}$	33,870	16,930	1470
10	32-10	E	5-3	$6\frac{11}{16}$	$11\frac{15}{16}$	38,370	19,180	1700
11	36-1	E	5-7	$6\frac{11}{16}$	$12\frac{7}{16}$	42,750	21,370	1930
12	39-4	E	5-11	$7\frac{1}{2}$	$13\frac{13}{16}$	47,130	23,560	2360
13	42-8	E	6-3	$7\frac{1}{2}$	$14\frac{5}{16}$	51,630	25,810	2640
14	45-11	E	6-7	$7\frac{1}{2}$	$14\frac{13}{16}$	56,000	28,000	2930
15	49-2	E	6-11	$7\frac{1}{2}$	$15\frac{3}{8}$	60,380	30,190	3220
16	52-6	E	7-3	$7\frac{1}{2}$	$15\frac{7}{8}$	64,880	32,440	3530
9	29-6	F	4-11	$6\frac{11}{16}$	$11\frac{7}{16}$	27,100	13,550	1460
10	32-10	F	5-3	$6\frac{11}{16}$	$11\frac{15}{16}$	30,700	15,350	1680
11	36-1	F	5-7	$6\frac{11}{16}$	$12\frac{7}{16}$	34,200	17,100	1920
12	39-4	F	5-11	$6\frac{11}{16}$	13	37,700	18,850	2160
13	42-8	F	6-3	$6\frac{11}{16}$	$13\frac{1}{2}$	41,300	20,650	2410
14	45-11	F	6-7	$6\frac{11}{16}$	$14\frac{1}{16}$	44,800	22,400	2680
15	49-2	F	6-11	$6\frac{11}{16}$	$14\frac{9}{16}$	48,300	24,150	2960
9	29-6	G	4-11	$6\frac{11}{16}$	$11\frac{7}{16}$	21,450	10,720	1460
10	32-10	G	5-3	$6\frac{11}{16}$	$11\frac{15}{16}$	24,300	12,150	1670
11	36-1	G	5-7	$6\frac{11}{16}$	$12\frac{7}{16}$	27,080	13,540	1900
12	39-4	G	5-11	$6\frac{11}{16}$	13	29,850	14,920	2150
13	42-8	G	6-3	$6\frac{11}{16}$	$13\frac{1}{2}$	32,700	16,350	2400
14	45-11	G	6-7	$6\frac{11}{16}$	$14\frac{1}{16}$	35,470	17,730	2660
15	49-2	G	6-11	$6\frac{11}{16}$	$14\frac{9}{16}$	38,240	19,120	2930

Courtesy Centrecon, Inc.

Table 7-4b. Dimensions and Strengths—Square Hollow Concrete Poles (Courtesy Centrecon, Inc.)

Overall pole length, ft	Pole size— tip/butt, in	E.P.A., ft² Concrete strength, lb/in²		Ultimate ground-line moment, ft·lb	Breaking strength load 2 ft. below tip, lb	Deflection per 100 ft·lb, in	Deflection limitations, ft·lb	Pole weight, lb
		6000, standard	7000, when specified					
25	7.6/11.65	43.3	45.3	84,000	4540	0.03	4100	2260
30	7.6/12.46	38.7	40.7	132,000	5740	0.03	5000	2880
35	7.6/13.27	33.8	36.2	147,000	5350	0.03	5900	3600
40	7.6/14.08	29.4	31.7	163,000	5090	0.03	6800	4370
45	7.6/14.89	25.7	28.1	178,000	4880	0.06	3850	5225
50	7.6/15.70	22.3	24.7	193,000	4710	0.06	4300	6160
55	7.6/16.51	19.1	21.7	209,000	4590	0.06	4750	7270

Glossary of Terms

E.P.A.   Effective projected area, in square feet of transformers, capacitors, streetlight fixtures, and other permanently attached items which are subject to wind loads.

Concrete strength   This is a reference to the compressive strength of the concrete in pounds per square inch as measured by testing representative samples 28 days after casting.

Ultimate ground-line bending moment   This is the bending moment applied to the pole which will cause structural failure of the pole. This is the result of multiplying the load indicated in the column Breaking Strength by a distance 2 ft less than the pole height (i.e., 2 ft less than the length of pole above ground). Figures under Ultimate Ground-Line Moment assume embedment of 10 percent of the pole length plus 2 ft. The figures in this column on technical charts are maximum moments expected to be applied to the pole. Appropriate safety factors should be used by the designer.

Breaking strength   This is the approximate load which, when applied at a point 2 ft below the tip of the pole, will cause structural failure of the pole.

Ground line   The point at which an embedded pole enters the ground or is otherwise restrained.

Deflection   The variation at the tip of the pole from a vertical line resulting from the application of loads such as equipment, wind, ice, etc.

Ground-line bending moment   The product of any load applied at any point on the pole multiplied by its height above ground line.

Dead loads   This refers to the load on a pole resulting from the attachment of transformers and other equipment permanently.

Live loads   These are loads applied to the pole as a result of wind, ice, or other loads of a temporary nature.

Metal Poles

Similar to concrete poles, metal poles of steel and aluminum are made in round, square and polygon shapes. They may also be formed from angles, channels or tees, and sometimes are laced together for greater strength. They may be set directly in the ground or, for larger sizes, bolted to a concrete base. Because of their appearance, they are installed in special instances in distribution systems and more widely used in transmission systems in place of steel towers and wood structures, particularly where the width of the right-of-way may be limited.

Towers

Steel towers are made up of angles and other shapes bolted or riveted together to form a rigid, strong and self-supporting structure. As such, they may be considered as another form of steel pole. The stresses imposed on them are determined in the same manner as for wood poles previously described. The towers are usually very tall as the conductors they support consist of long spans. The long spans impose the additional problems of relatively large stresses that may be imposed on a tower from broken conductors and from vibrations resulting from galloping (or dancing) conductors caused by wind, sleet and ice. In determining the stresses associated with the several members of the steel structure, not only are the tension and compressive yield point values to be considered, but also the shear and bearing values of the bolts and rivets involved. The working load stresses used in the design of the towers may be a percentage of the yield values determined above, the percentage depending on the importance and type of the line.

For economic reasons, most towers are designed to support two or four circuits, one or two on each side, spaced far enough apart so that the conductors (in a long span) do not hit each other in a wind or when galloping, and more especially if one or more conductors break; crossarms must be designed to accommodate the conductors.

The factors that affect the design of towers, their shape and strength include: type of tower, single or double circuit; height of tower (fixed by the sag, span, length of insulator string and distance between conductors and the ground); permissible distance between phases; permissible distance between circuits; minimum distance between conductors and members of the tower structure. This last may be the length from point of support equal to the length of the insulator string, swung at the angle of maximum transverse deflection.

The conductor size is usually selected to meet electrical requirements economically. The maximum stress in the conductor can, however, be controlled by the tension at which the conductor is strung.

The span length is determined generally by economic considerations in which insulator costs may play an important part.

The type of tower (Figure 7-4a, b, and c), whether so-called line or suspension type, or dead-end tower, depends on the "straightness" of the line. For straight portions of the line, or where only relatively small angles (say to 20°) exist, the spans on both sides of the insulator string tend to balance each other, so that the conductors are attached to a string of insulators that hangs essentially perpendicularly. For greater angles or for dead-end towers, the conductor may be supported by one or more strings (in parallel) of insulators that hang more or less horizontally. In this instance, when a conductor breaks, the insulator strings swing into a catenary, increasing its length, greatly reducing the tension in the span adjacent to the break; the reduction may be as great as 25 to 40 percent. Generally, dead-end towers are designed to resist all conductors broken at maximum stress in the conductor. To protect a long line from the "domino effect" of broken conductors or tower failure, dead-end towers (and poles) are inserted strategically in points along the straight line.

Transpositions of conductors of a circuit are made to reduce the overall inductive effect between conductors of the same circuit, adjacent circuits and communication lines. Transpositions are made on structures with special attachments or on structures specially built for that purpose.

Like pole lines, where the stress on the structure exceeds the ability to withstand them, guying is employed to furnish the additional strength needed.

Tower construction must provide protection from lightning. As described earlier, a ground or shield wire installed above the transmission circuit is effective if placed to provide a 30° angle of shielding over the conductor, Figure 7-4d. While this can be attained in the case of single-circuit towers and steel-pole lines, it is easier of attainment for double-circuit towers to provide for two shield or ground wires, each placed approximately over a circuit. This may necessitate a crossarm to be constructed at the top of the tower. Coupled with the ground wire is the grounding of the tower footings, for a low resistance ground is essential to the effectiveness of the shield wire. Ordinarily, the tower footings are grounded by connection to a number of ground rods driven into the earth. When the ground resistance is high, it may be necessary to bury conductors radiating out from the tower footings. A buried conductor constitutes a counterpoise. Other counterpoises may be more elaborate and may be a combination of the various schemes, Figure 7-6.

In addition to the metal poles and towers for supporting the conductors, there are other types of structures, named appropriately after their shapes and appearance, Figure 7-5. They are referred to as A frames, H frames, V frames, and Y frames; the first three may be constructed of wood or metal (steel or aluminum), the last or Y is usually only made of steel. Practically all of these structures require guying for stability and to achieve their strength. Also, the crossarms must be such as to accommodate the string insulator configuration to prevent contact between conductors or between conductors and the structure during periods of maximum sway.

Figure 7-4c. Dead-end tower.

GUYS

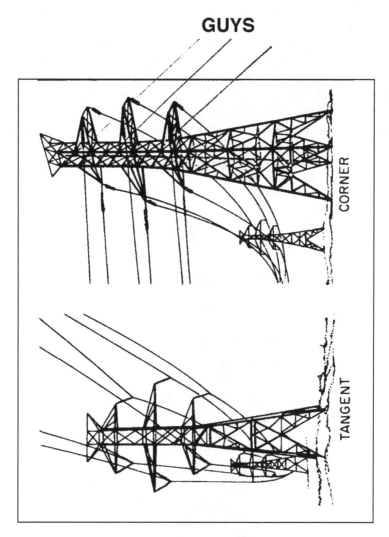

Figure 7-4b. Angle or corner tower.

Figure 7-4a. Tangent or suspension tower constructions.

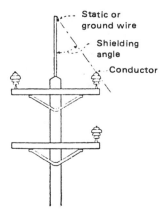

Figure 7-4d. Overhead Ground or Static Wire to Protect Transmission Line

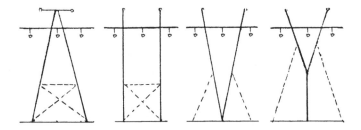

Figure 7-5. Transmission Line Supporting Structures

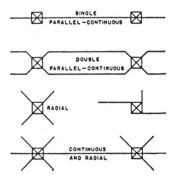

Figure 7-6. Arrangements of Counterpoise
(Courtesy Westinghouse Electric Co.)

Ground resistance may be measured between the tower or structure connected to the counterpoise and electrodes driven into the earth a known distance away, Figure 7-7. The measurements may vary with moisture, temperature, season of the year, earth composition and pollution, as well as the depth and diameter of the electrodes employed.

Ground resistance may be reduced by installation and connection of additional ground rods, by adding conductors to the counterpoise, or both. Some soil resistivity values are indicated in Table 7-5. Similar

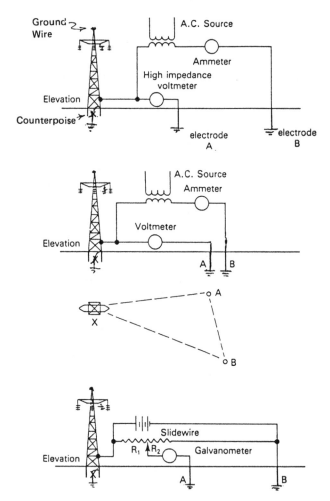

Figure 7-7. Measuring Ground Resistance Methods (*Courtesy Westinghouse Electric Co.*)

methods may also be applied to substation structures.

Table 7-5. Typical Values of Soil Resistivity*
(Courtesy Long Island Lighting Co.)

Soil	Resistivity Range
Clay, moist	14-30
Swampy ground	10-100
Humus and loam	30-50
Sand below ground water level	60-130
Sandstone	120-70,000
Broken stone mixed with loam	200-350
Limestone	200-4,000
Dry earth	1,000-4,000
Denserock	5,000-10,000
Chemically pure water	250,000
Tap water	1,000-12,000
Rain water	800
Sea water	0.01-1.0
Polluted river water	1-5

*In ohms per cubic meter.

RIGHTS-OF-WAY

Generally speaking, poles require less width of right-of-way than do tower lines, and towers require less than the A, H, V and Y structures. All, however, are subject to NESC recommendations and local and environmental requirements.

Overhead transmission lines are particularly suited for open country areas where the line may be reasonably straight and amply wide rights-of-way are not only available, but relatively easy to acquire; further, appearances in such cases are not always important. National and local safety code clearances for the voltage ranges involved make such open country installations economically and environmentally acceptable.

Where transmission lines must pass through populated urban and suburban areas, relatively narrow streets and back alleys furnish the rights-of-way. Here, the tall metal poles, usually carrying a single circuit are employed with span lengths limited to a city block or less. Guying, tree conditions and other factors must conform to acceptable appearance standards. Included also may be limited access in which materiel may have to be carried by hand and limits placed on working hours.

Clearing of trees, brush and other growth from rights-of-way must be maintained so that future growth will not interfere with operation of the lines. In wooded areas, trees must be cleared or topped far enough from the right-of-way so that falling trees will not inflict damage to the lines. Typical specifications are illustrated in Figure 7-8.

Access roads must be provided for initial construction and future inspections, patrols, etc.; these and the right-of-way must be cleared so that vehicles may travel unimpeded. In some instances, helicopters may be employed to deliver personnel and material, sometimes preassembled, to the job site for construction and maintenance.

Utilization of railroad right-of-way may provide a desirable location for transmission lines, especially as the railroads serve population and industrial centers that also constitute electrical load centers. There are some drawbacks, however. Soot and smoke from coal- and oil-fired locomotives accumulate on the surface of insulators that must be periodically cleaned to prevent flashover, and protracted delays may be experienced in scheduling work to comply with railroad operations. Further, the high voltage of the line may cause interference in communication circuits that often occupy the same right-of-way.

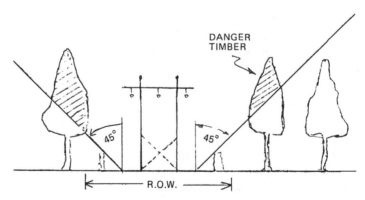

Figure 7-8. Clearing of Right-of-Way

ELECTROMAGNETIC INDUCTION

The effect on human, animal and plant life of the magnetic fields from high voltage transmission lines in their vicinity is the subject of continuing research. Fundamentally, when a moving magnetic field and a conductor cross each other, a voltage is induced in the conductor depending in part on the strength of the magnetic field. People, animals and plants are conductors and the voltage induced in them causes eddy currents to flow within them, between them and the ground, and between them and other objects with which they may come into contact.

Although the effect of such currents on the biological, and especially on the nervous system appears minimal, the duration of the exposure may play some part. Meanwhile some utilities and government agencies have developed tentative and precautionary codes for minimum width of right-of-way of some 350 feet or 170 meters with additional width required to maintain a maximum magnetic field strength of 1.6 kV per meter from conductor to the area or structure in the vicinity applied to the shortest distance between them. These are minimum specifications and may be modified to meet local and particular situations. They apply to both alternating and direct current high voltage transmission line rights-of-way.

CROSSARMS

Crossarms are generally used to carry polyphase circuits. They are also used where lines cross each other or make sharp turns at large angles to each other. Alley or side arms are used in narrow rights-of-way, the greater part of the arm extending out on one side of the pole. Where appearance is important, other means may be employed in place of the crossarm.

The crossarm is essentially a beam supported at the point of attachment to the pole, Figure 7-9 It must support the vertical loadings from the weight of ice-covered conductors and, for safety reasons, the weight of workers, some 200 to 250 pounds. It is also subject to horizontal loadings of winds, tension in conductors where those on each side of the crossarm do not balance each other (unlike spans or conductors, deadend, bends, offsets, etc.) and from possible conductor breakage.

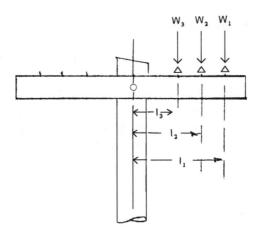

Figure 7-9. Bending Moments on Crossarm

Stresses

In determining stresses, the same principles for determining beam stresses are employed as in the case of poles, previously described.

The total bending moment M is equal to the sum of all the individual loads multiplied by their distances from the cross section under consideration. The weakest section usually should be at the center of the crossarm where it is attached to the pole. The cross section at the pin holes, however, is reduced and may be the weakest point in the crossarm. The determination can be made by calculating unit fiber stress at the pin location:

$$f = \frac{M}{I/c}$$

where   f = maximum unit fiber stress at extreme edge of the crossarm, in lbs/in.^{2}
  M = total bending moment, in inch-pounds
  I = moment of inertia of cross section
  c = distance from neutral axis to extreme edge, in inches

The moment of inertia for a rectangular cross section is:

$$I = \frac{1}{12}bd^2 \qquad \text{and} \qquad c = \frac{d}{2}$$

and the section modulus:

$$\frac{I}{c} = \frac{1}{6}bd^2$$

where the neutral axis is parallel to side d (Figure 7-10c).

Where the cross section is lessened by the pin hole in the crossarm, the section modulus becomes:

$$\frac{I}{c} = \frac{1}{6}d\left(b^2 - \frac{a^3}{b}\right)$$

where a is the diameter of the hole.

Where stresses on a crossarm approach or exceed safe values (depending on the kind of wood), resort is had to double arms, that is, a second crossarm is mounted on the other side of the pole and bound together by bolts and spacers of wood or steel. Each crossarm is usually fastened to the pole and steadied in position by flat braces, usually of steel but sometimes of wood. Where these measures are still insufficient, preformed steel angles are employed, and if still greater strength is required, the crossarms are guyed to adjacent poles, Figure 7-11.

The strength and stability of the crossarm are also dependent on

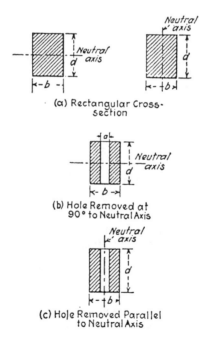

(a) Rectangular Cross-
Section

(b) Hole Removed at
90° to Neutral Axis

(c) Hole Removed Parallel
to Neutral Axis

Figure 7-10. Cross Sections of Crossarms
(From Overhead Systems Reference Book)

the bolt through which the stresses are transferred to the pole. The pressure on the bolt in the crossarm is:

$$P_A = \frac{W}{b_A d}$$

and that on the pole is:

$$P_p = \frac{W}{b_p d}$$

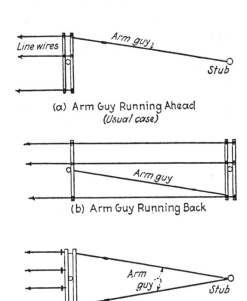

(a) Arm Guy Running Ahead
(Usual case)

(b) Arm Guy Running Back

(c) Double Arm Guys *(Heavy loading)*

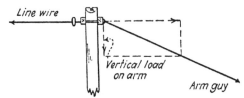

(d) Vertical Loading on Arm due to Arm Guy

Figure 7-11. Arm Guys
(From Overhead Systems Reference Book)

where W is the weight of the load on the crossarm
        bA is the width of the crossarm
        b p is the diameter of the pole
and   d is the diameter of the bolt

The maximum unit pressure must not exceed the bearing value of the wood or distortion takes place. As the ultimate strength is approached, the bolt tends to bend and the fibers of the wood begin to give way, Figure 7-12.

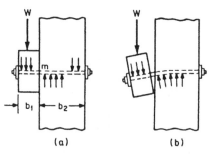

Figure 7-12. Action on Bolt Holding Crossarm to Pole m **is Point of Maximum Shear Stress** *(From Overhead Systems Reference Book)*

Steel Crossarms

Where stresses exceed even the double-arm capability, steel crossarms may be used. Stresses are computed in much the same way as for wood. The steel crossarm does not have the insulating value of a wood crossarm and is much heavier to handle.

PINS

Pins support the conductors with their ice coatings and are subject to both vertical and horizontal loadings from dead weight and wind pressure, and tensions, similar to pole loadings. Under vertical load, the pin acts as a column transmitting its load to the crossarm and pole. Compared to the horizontal loading, this value is small and usually neglected. The horizontal loading acts on it as a beam, Figure 7-13, and the bending moment M is:

$$M = Ph$$

where P is the load on the pin and h the distance from the conductor to the base of the pin. The maximum fiber stress is usually where the shoulder of the pin contacts the crossarm. Its unit value in pounds per square inch is:

$$f = \frac{Ph}{0.0982d^3}$$

where d is the diameter of the shank of the pin.

Like poles, the weak point is about one-third of the distance down from the conductor.

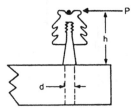

Figure 7-13. Loading on Pins
(From Overhead Systems Reference Book)

Where stresses are great, as with crossarms, double pins are employed, one on each of the double arms.

With improvement in materials and work methods, together with more emphasis on appearance, so-called armless construction is used. Here, steel pins mounted directly on the pole take the place of crossarms and pins, as shown in Figure 7-14. Vertical and horizontal loadings are as indicated in the diagram. Stresses on the pin are calculated in the same manner as for crossarms and pins.

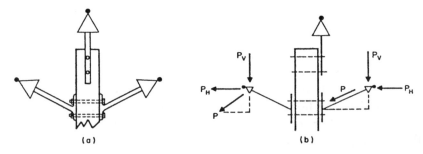

Figure 7-14. Pins in Lieu of Crossarm Construction. (a) Use of Long Steel Pins. (b) Stresses on Long Steel Pins. *(Courtesy Ohio Brass Co.)*

For higher distribution voltages (up to 45 kV) and lower transmission voltages (69 kV and below), similar construction is employed for polyphase circuits using post-type insulators mounted at right angles to the pole, either all on one side or alternating on both sides of the pole.

Secondary mains (and services) employing cabled conductors are mounted directly to the pole using clamps supporting the neutral conductor around which the conductors are cabled. The vertical and horizontal loadings are transmitted to the pole by the bolts that attach the clamps to the pole.

Secondary mains in a great number of installations are attached to the pole by means of secondary racks. The conductors are attached to spool-type insulators on a common shaft, the whole attached to a steel backing bolted to the pole. Vertical and horizontal stresses are determined in a manner similar to that for poles, crossarms, etc.

INSULATORS

Insulators are most commonly made of porcelain, although glass insulators exist in relatively great numbers on the older 2.4/4.16 kV primary and 120/240 volt secondary lines. Porcelain has little tensile strength but great compressive strength (as does glass) and, hence, lines are so designed that insulators will be in compression in carrying the mechanical loads imposed on them. Several types of insulators are described below.

Pin Type

The strength of pin type in compression is usually greater than of the pin upon which it is mounted. The physical dimensions of the insulators necessary to meet the mechanical requirements are usually sufficient in meeting the electrical requirements when wet, including surge voltages.

Post Type

Post-type insulators are essentially pin-type insulators that incorporate their own steel pins. As mentioned earlier, this type insulator may be installed in a vertical, or near vertical position as well as horizontally, or nearly so, in place of crossarms to carry the conductors of a polyphase circuit. The horizontal loadings create stresses of compression on one side of the pole and tension on the other, both transmitted to the pole through the steel pin. The vertical loadings result in a stress of compression in the porcelain between the conductor and the steel pin, the latter transmitting the stress to the pole. Where the post insulator is mounted at an angle to the pole, the stress will consist of the component of the horizontal and vertical forces acting on the pin and the porcelain.

Suspension or Strain Type

These are also referred to as disc or string insulators and are almost exclusively used on transmission lines where stresses are usually greater than those associated with distribution systems. The number of discs strung together depend on the operating voltage of the line.

Strain or Ball Type

This type has been used to dead-end lower voltage primary and secondary conductors, and as insulators in guy wires in older installations, many of which still exist. Here, the porcelain is in compression between the stresses imposed by the forces acting in the guy wire.

Spool Type

These are almost always used only with secondary racks. The compressive strength of the porcelain here is usually greater than the strength of the other parts of the rack.

Other Types

Insulators of the knob type are sometimes used for services and on secondary mains. Other types include bushings and bus supports.

GUYS AND ANCHORS

Where horizontal loads are imposed on poles and crossarms, guys are used to take up the horizontal stress and transmit it to other poles, crossarms, or into the ground. The various types of guys are illustrated in Figure 7-15. Where guys cannot be installed because of space limitations, crib bracing is used to provide additional holding power, but does not add to the strength of the pole.

On long straight lines, with few side taps, guys are installed at right angles to the line at strategic locations. Their purpose is to mini-

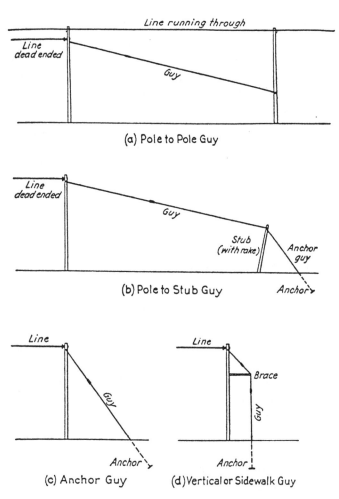

Figure 7-15 Pole Guys
(From Overhead Systems Reference Book)

mize damage caused by severe storms or accidents that can result in a "domino effect" of collapsing pole line. They are sometimes referred to as storm guys.

Loading

Guys have loads imposed on them from tension in the conductors and the angle between adjacent conductor spans. The magnitude of the tension is based on the conductor sizes, including ice and wind loads, and the sag in the span. Design limits are based on the elastic limits of the conductors, usually 50 to 60 percent of the ultimate strength of the conductor metal. As the design limits are based on the worst loading conditions that happen occasionally, the usual stress is generally less than 50 to 60 percent of the elastic limit.

Poles at an angle in the line undergo stresses also due to the tension in the conductors, but the guy handles only a component of that tension, the amount depending on the angle in the line, Figure 7-16.

If T is the sum of all the tensions caused by the conductors and the angle of the line is a, the component of T in line with the guy is:

$$T_a = T \sin \frac{a}{2}$$

and the total stress the guy handles is twice that:

$$T_{guy} = 2T \sin \frac{a}{2}$$

The stress handled by the guy will be the vector sum of the tensions in the two spans that are not balanced. If the angle between spans is relatively large, usually more than 60°, the load on the guy that bisects the angle will be greater than the dead-end loading of the line and, if practical, two dead-end guys are preferable.

As near as practical, the guy should be attached to the center of loading of the loads it supports. When the loads are at different points on the pole, they should be converted into a single equivalent load at the point of attachment.

If T_p is the loading at height h_p
    T_s is the loading at height h_s
    P_w is the wind pressure on the pole concentrated at h_w
    L_H is the horizontal equivalent loading

Then $L_H = \dfrac{T_p h_p + T_s h_s + P_w h_w}{h}$

As generally the guy is not horizontal, the actual tension in it will be greater than L_H. If b is the angle the guy makes with the horizontal, then

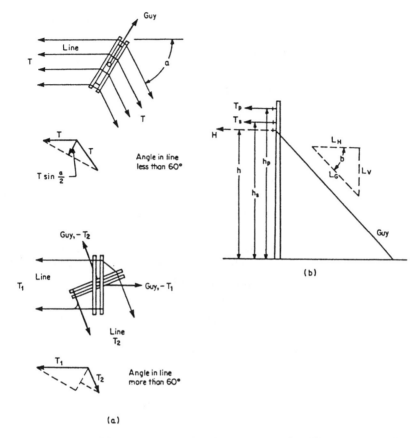

Figure 7-16. Loading on Guys. (a) Guys at Angles (b) Guys Loading.
(From overhead Systems Reference Book)

the loading in the guy becomes:

$$L_G = \frac{L_H}{\cos b}$$

and the vertical component is:

$$L_V = L_H \tan b \text{ or } L_V + L_G \sin b$$

and is an additional vertical load on the pole.

If the guy is attached too far from the center of the load the pole section above that point acts as a beam, and the moment then will be:

$$M = T_p(h_p - h) + T_s(h_s - h) - P_w(h - h_w)$$

and the fiber stress at the point of attachment will be:

$$f = \frac{M}{0.0982 \, d^3}$$

where d is the diameter of the pole at that point.

The guy should be so attached as near as practical at the center of the load so that it takes the entire horizontal load with the pole acting as a strut.

Guy wires come in many sizes but, for practical purposes, they may be limited to four sizes, whose characteristics are given in Table 7-6. If stresses exceed the maximum strength of one of the wire sizes, the next larger size or two guys should be used.

Guy wires are attached to poles and crossarms by eye bolts, clamps, thimbles, clips and plates, and by special eye-shaped ends bent at an angle to accommodate the guy wire.

Table 7-6. Guy Wire Characteristics
(Courtesy Long Island Lighting Co.)

Wire Class	Ultimate strength, lb/in²	Elastic limit lb/in²
Standard	47,000	24,000
Regular	75,000	38,000
High-strength	125,000	69,000
Extra high-strength	187,000	112,000

Weight of steel wire: 0.002671 lb/in³
Modulus of elasticity: 29 × 10⁶
Coefficient of linear expansion: 11.8 × 10⁻⁶/°C; 6.62 × 10⁻⁶/°F
Within the 1/4- to 1-2-inch range, for the four classes of wire, ultimate strengths
    vary from a minimum of 1900 lb to a maximum of 27,000 lb.

Push Braces

Push braces are sometimes installed where guys are impractical to install. They are essentially compression type "guys" where the brace pole takes the place of the guy wire. Stresses are determined in the same manner as for wire guys.

Anchors

Obviously, the holding power of the anchor should match the strength of the associated guy wire. In general, the holding power depends on the area the anchor offers the soil and the depth at which it is buried and the nature of the soil, that is, the weight of the soil constituting the resisting force. Types of anchors, classification of soils and the selection of anchors are shown in Figure 7-17 and Tables 7-7 and 7-8.

CONDUCTORS

Tensions and Sag

In addition to the problems of ice and wind affecting the stringing of conductors, there is the problem of how tight the conductors should be strung. If stretched too tightly, the stresses imposed on the pole and its appurtenances (crossarms, pins, insulators, racks, and hardware) would be so great as to make the arrangement impractical. The stresses on the conductors themselves may cause them to exceed their elastic limits, should the structure move even slightly. The resulting elongation may become permanent with a reduction in the cross section of the conductor, leading to possible failure.

If the conductor is stretched too loosely, the resulting increase in sag would affect the swaying of conductors that might necessitate wider spacing in both the horizontal and vertical planes.

Proper sagging of the conductor would eliminate both of these possibilities. The tension in a conductor is determined by the sag, being inversely proportional to the sag. In determining the final sag, not only are loading conditions considered, but also the probable temperature variations and local physical conditions as well as regulations and code restrictions.

In determining tensions and sags in a conductor, it should be assumed that the loading is uniformly applied over its length with the conductor freely shaping itself into a catenary. For relatively short spans,

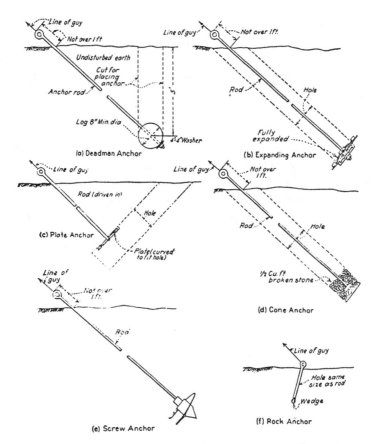

Types of Anchors. Anchors come in many shapes and types. They may be classified into four general types:

1. Buried logs, planks, or plates attached to the end of a rod.
2. Screw anchors, screwed into the soil at varying depths. A very large screw anchor, known as the swamp anchor, is used in swampy areas.
3. Expanding anchors, in which a plate in sections is folded into a small diameter, the unit set into a small-diameter hole (or at the bottom of a pole), and the anchor rod screwed or pounded into it so that the sections spread out, biting into the adjacent soil. If the expanding anchor plate is divided into eight sections, for example, the anchor is known as an eight-way expanding anchor.
4. Rock anchors, which are merely rods driven into the rock, hard shale, or hardpan, at approximately right angles to the guy wire. The depth at which they are installed will vary with the strength required and the character of the rock.

Figure 7-17. Types of Anchors
(From Overhead Systems Reference Book)

Table 7-7. Classification of Soils

Class	Description
1	Hard rock: solid.
2	Shale, sandstone: solid or in adjacent layers.
3	Hard, dry: hardpan, usually found under class 4 strata.
4	Crumbly, damp: clay usually predominating. Insufficiently moist to pack into a ball when squeezed by hand.
5	Firm, moist: clay usually predominating with other soils commonly present. Sufficiently moist to pack into a firm ball when squeezed by hand (most soils in well-drained areas fall into this classification).
6	Plastic, wet: clay usually predominating as in class 5, but because of unfavorable moisture conditions, such as in areas subjected to seasonally heavy rainfall, sufficient water is present to penetrate the soil to appreciable depths and, though the area be fairly well drained, the soil becomes plastic during such seasons, and when squeezed will readily assume any shape (a soil not uncommon in fairly flat areas).
7a	Loose, dry: found in arid regions, sand or gravel usually predominating (filled-in or built-up areas in dry regions fall into this class, and as the name implies, there is very little bond to hold the particles together).
7b	Loose, wet: same as loose, dry for holding power; high in sand, gravel or loam content. Holding power in some seasons is good, but during rainy seasons soil absorbs excessive moisture readily with resultant loss of holding power, especially in poorly drained areas. This class also includes soft wet clay.
8	Swamps and marshes.

Courtesy Long Island Lighting Co.

Table 7-8. Selection of Anchors-Approximate Holding Power, lb
Diameter of Screw or Expanded Anchor Plate; Diameter and Length of Rod

	Screw*	Expanding			Swamp**	
		Eight-way	Eight-way	Four-way		
	8-in	8-in	10-in	12-in	13-in	15-in
Soil class	1 in × 5.5 ft	3/4 in × 8 ft	1 in × 10 ft	1-1/4 in × 10 ft	11 2-in pipe	2-in pipe
1***	NR	NR	NR	NR	NR	NR
2***	NR	NR	NR	NR	NR	NR
3	NR	26,500	31,000	40,000	NR	NR
4	11,000	22,000	26,500	34,000	NR	NR
5	8,000	18,500	21,000	26,500	NR	NR
6	6,500	15,000	16,500	21,500	NR	NR
7	3,500	10,000	12,000	16,000	NR	NR
8	NR	NR	NR	NR	12,000	15,000

*Screw anchors are used especially in temporary installations because of easy removal.
** At least one 10-ft length of 2-in pipe should be installed; additional lengths should be installed until pipe can no longer be turned (say, by four workers operating the wrenches).
***Special rock anchors of varying holding power should be used.
NR—Not recommended
Courtesy Long Island Lighting Co.

as in distribution systems, the error is small and within practical field construction practices and may be neglected.

For parabolas, the relation between sag or deflection (d), tension (T) and span length (L) is, Figure 7-18.

$$d = \frac{wL^2}{8T}$$

where d and L are in feet, T in pounds, and w the resultant load (conductor, ice, etc.) in pounds per foot.

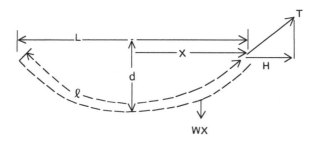

Figure 7-18. Loading on a Conductor

For long spans, the horizontal component H of the tension in the conductor is (by right-angle triangle relation):

$$H^2 = T^2 - (wx)^2$$

where x is one-half the length of the conductor in the span, or of the span itself. The conductor at the support where the tension T is maximum, is at an angle $\tan^{-1}(wx/H)$ to the horizontal. Its length ℓ compared to the span length L is:

$$\ell = L + \frac{8d^2}{3L}$$

where d is the sag in feet.

With temperature increase, the conductor expands in length, the sag increases and the tension in the conductor decreases. The elongation from the *change* (t) in temperature also depends on the coefficient of expansion (e) and its length (ℓ):

$$\text{Elongation} = \ell \text{ et}$$

The elongation of a conductor decreases when the tension is decreased. If the loading on the conductor is increased, the tension and accompanying elongation are increased, in turn increasing the sag and decreasing the tension. The elongation or change length becomes:

$$\text{Change in length} = \frac{T_s}{aE}$$

where $T^1 - T^2$ is the change in tension, a the cross section area, and E the modulus of elasticity.

For longer transmission spans, the difference may be significant and mathematical methods of calculating catenaries may be used. This discussion may be omitted, if desired.

The Catenary

The load of the conductor with a coating of ice is assumed to be uniformly distributed along its entire length; refer to Figure 7-19.

Let   w = the load per unit along the conductor
     ℓ = the length of the conductor between supports
   wℓ = the total load on the entire conductor

A portion of the conductor from the lowest point 0 to any other point B, is shown in Figure 7-19.

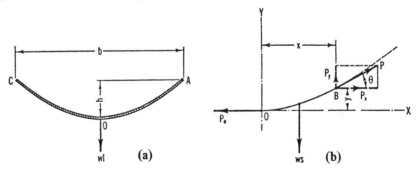

Figure 7-19. Catenary

and s = length of the portion of the conductor
         P_0 = tensile force at point 0, in equilibrium
         P = tensile force at point B
         ws = load of the portion of the conductor

then $\Sigma F_x = 0$ or $P_x - P_0 = 0$ and $P_x = P_0$

         $\Sigma F_x = 0$ or $P_y - ws = 0$ and $P_y = Ws$

$$\tan \theta = \frac{P_y}{P_x} = \frac{ws}{P_0}$$

Since P is tangent to the curve at point B

$$\tan \theta = \frac{d_y}{d_x} \text{ and } \frac{d_y}{d_x} = \frac{ws}{P_0}$$

(calculus) $ds^2 = dx^2 + dy^2$

dividing by dy^2

$$\left(\frac{ds}{dy}\right)^2 = \left(\frac{ds}{dy}\right)^2 + 1$$

substituting $\left(\frac{ds}{dy}\right)^2 = \left(\frac{P_0}{ws}\right)^2 + 1$

and

$$dy = w\frac{s\,ds}{\sqrt{P_0^2 + w^2 s^2}}$$

$$y = w\int_0^s \frac{s\,ds}{\sqrt{P_0^2 + w^2 s^2}}$$

$$y = \frac{\sqrt{P_0^2 + w^2 s^2}}{w} - \frac{P_0}{w}$$

that is, the sag at point B.
    Going back to equation $ds^2 = dx^2 + dy^2$

$$\left(\frac{ds}{dy}\right)^2 = \left(\frac{ds}{dy}\right)^2 + 1$$

substituting $\qquad \left(\frac{ds}{dy}\right)^2 = \left(\frac{P_0}{ws}\right)^2 + 1$

and

$$dx = P_0 \frac{ds}{\sqrt{P_0^2 + w^2s^2}}$$

$$x = P_0 \int_0^s \frac{ds}{\sqrt{P_0^2 + w^2s^2}}$$

$$x = \frac{P_0}{w}\left[\log_e\left(ws + P_0^2 + w^2s^2\right) - \log_e P_0\right]$$

$$\frac{wx}{P_0} = \log_e \frac{ws + \sqrt{P_0^2 + w^2s^2}}{P_0}$$

$$e^{wx/P_0} = \frac{ws + \sqrt{P_0^2 + w^2s^2}}{P_0}$$

$$s = \frac{P_0}{2w}\left(e^{wx/P_0} - e^{-wx/P_0}\right)$$

substituting this value of s in the equation for y above

$$y = \frac{P_0}{w}\left[\frac{1}{2}e^{wx/P_0} + e^{-wx/P_0} - 1\right]$$

Tension at Lowest Point

Referring to Figure 7-19a, the quantities b/2 and h are the coordinates at point of support A. Substituting $x = b/2$ and $y = h$ in the above equation for y:

$$h = \frac{P_0}{w}\left[\frac{1}{2}\left(e^{wb/2P_0} + e^{-wb/2P_0}\right) - 1\right]$$

Tension at Any Point

From Figure 7-19b:

$$P = \sqrt{P_x{}^2 + P_y{}^2}$$

Substituting the values of P_x and P_y from the equation above,

$$P = \sqrt{P_x{}^2 + w^2 s^2}$$

Substituting the value of s from the equation above,

$$P = P_0 \sqrt{1 + 1/4\left(e^{wx/P_0} - e^{-wx/P_0}\right)^2}$$

Substituting the value of s from the equation above,

$$\tan \theta = 1/2\left(e^{wx/P_0} - e^{wx/P_0}\right)$$

Maximum Tension in the Conductor

From the above equation, P is a maximum when x is a maximum. The maximum value of $b/2$, indicating that the tension is greatest at the point of support, P_A. Substituting $x = b/2$ in the equation above:

$$P_A = P_0 \sqrt{1 + 1/4\left(e^{wb/2P_0} - e^{-wb/2P_0}\right)^2}$$

and
$$\tan \theta_A = 1/2\left(e^{wb/2P_0} - e^{-wb/P_0}\right)$$

Length of Conductor (ℓ)

Substituting $x = b/2$ in the equation above and multiplying by 2:

$$\ell = \frac{P_0}{w}\left(e^{wb/2P_0} - e^{-wb/2P_0}\right)$$

The quantities involved in the several formulas above do not lend themselves readily to algebraic methods of solution. For practical purposes, trial methods in which various values are substituted for the unknown quantities can be used until a value is found that closely satisfies the equation. Logarithmic tables simplify the procedures.

The above formulas may be expressed and calculations expedited by the use of hyperbolic functions. The expression $1/2(e^x - e^{-x})$ is the hyperbolic sine of x and is written sinh x; $1/2(e^x + e^{-x})$ is the hyperbolic cosine of x and is written cosh x. Hence the expression

$$1/2 \ (e^{wx/P_0} + e^{-wx/P_0})$$

may be written cosh wx/P_0. The relation

$$\cosh^2 x - \sinh^2 x = 1$$

simplifies some of the formulas; the value of e is 2.7183.

Summarizing:

Sag
$$y = \frac{P_0}{w}\left(\cosh\frac{wx}{P_0} - 1\right)$$

Tension at lowest point
$$h = \frac{P_0}{w}\left(\cosh\frac{wb}{2P_0} - 1\right)$$

Tension at any point
$$P = P_0\cosh\frac{wx}{P_0}$$

$$\tan\theta = \sinh\frac{wx}{P_0}$$

Maximum tension in conductor
$$P_A = P_0\cosh\frac{wb}{2P_0}$$

$$\tan\theta_A = \sinh\frac{wb}{2P_0}$$

Length of conductor (ℓ)
$$= \frac{2P_0}{w}\sinh\frac{wb}{2P_0}$$

Various diagrams and curves have been devised to simplify the solution to the problems outlined above. These, and the tables of hyperbolic functions may be found in analyses that explore more extensively the application of mechanics to wires in suspension.

Spans Between Different Elevations

Spans between different elevations must be sagged so that the low point of the spans are below the elevations of the lower supports. If the low point of a span is higher, there will be an uplift at the lower elevation support. As loadings and temperatures change, the low point will move along the span in a horizontal line. For design purposes, however, this low point may be assumed to be fixed. From Figure 7-20, the approximate location of the low point may be determined:

$$x_1 = \frac{S}{2} + \frac{ht}{Sw} = S \frac{\sqrt{d}}{\sqrt{d-h} + \sqrt{d}}$$

$$x_2 = S - x_1 = \frac{S}{2}\left(1 - \frac{h}{4d_1}\right)$$

and
$$d_2 = d_1\left(1 - \frac{h}{4d_1}\right)^2$$

The horizontal components of the tension $t_1 = t_2$ and the vertical component at the upper support must be greater, that is,

$$t_1\ T_2 \text{ and } T_1 = T_2 \text{ -+ wh}$$

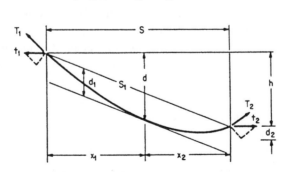

Figure 7-20. Span with Supports at Different Elevations

where w is the weight per foot of the conductor.

For practical purposes, in sagging the conductor in the field, it may be convenient to determine the sag as the vertical deflection from a line through the points of support. The sag may be computed as if the supports were at the same elevation and S the span length and measured as the vertical distance d_2 from the line through the points of support.

Conductor Materials

Overhead conductors must have low electrical resistance yet be economical; they must be strong so that mechanical failure be minimized as much as possible yet be workable; they should have a relatively small sag yet, because sag is approximately inverse to the tension, a large sag is desirable to hold stress as low as practical. This condition may best be met by conductors made of copper or aluminum, or in combination with steel, and a sag that will stress the conductor nearly to its elastic limit under the heaviest loading it may have to carry.

Copper wire is manufactured in three kinds: hard drawn, medium hard drawn, and soft drawn (annealed). Hard drawn is the strongest but least flexible that makes it relatively difficult to handle, while soft drawn is the weakest but easy to work with; medium hard drawn lies in between these two both in strength and ease in handling. The first is generally used for long transmission spans and some of the longer distribution feeder spans. The last, soft drawn, is usually limited to short distribution spans, services, and tying conductors to pin-type insulators. Medium hard drawn is used almost exclusively for relatively long spans in general use on distribution circuits.

Aluminum competes economically with copper for electrical conductors, even though its conductivity is only 63 percent as great; this is offset by its lighter weight, about one-third that of copper. For conductors of the same conductivity, aluminum is about half as heavy although of somewhat larger diameter. Aluminum is relatively low in tensile strength, being about two-thirds that of soft drawn copper, but because of the equivalent larger diameter, their tensile strengths are about the same. The larger diameter, in turn, results in greater ice and wind loads imposed on aluminum conductors. Data on these conductors are given in Tables 7-9 and 7-10.

To remedy these deficiencies, aluminum conductors are wrapped around or clad around steel wires: the aluminum gives the conductor its conductivity and the steel its mechanical strength. It is, therefore, considerably stronger than even hard drawn copper wire of equivalent conductivity. When strands of aluminum are wrapped around strands of steel, the conductor is referred to as Aluminum Conductor Steel Reinforced, or ACSR.

Sag for ACSR

For ACSR, the stresses on the steel and aluminum strands produce different results, each calculated in the same manner described earlier for conductors of a single material. For larger loads, the aluminum and steel act essentially as a single conductor. For lower value tensions, the strands tend to separate and the steel strands carry all of the load. The same action essentially results with temperature changes. The coefficient of expansion may be found, for practical purposes, from the coefficient of expansion a and the modulus of elasticity E and the percent area H for each of the metals involved:

$$E_{AS} = E_A H_A + E_S H_S$$

and
$$a_{AS} = \frac{E_A H_A}{E_{AS}} + a_S \frac{E_S H_S}{E_{AS}}$$

As solid conductors become larger, their rigidity increases and they are harder to handle. To remedy this condition, larger size conductors are usually stranded.

Steel wire is rarely used alone, usually for very long spans (such as river crossings), because of its 3 to 5 times greater strength, even though its conductivity is only one-tenth that of copper. Its tendency to rust may be counteracted by galvanizing or coating it with zinc.

Aluminum clad, as well as copper clad, steel wire is not only economical, but may be used as an electrical conductor where loads are relatively light, as in rural lines, and mechanically as guy wire.

Special precautions should be taken when conductors of copper and aluminum are connected together. Aluminum connectors with copper bushings are sometimes employed for this purpose.

Copper conductors on distribution circuits are sometimes covered with insulation of polyethylene (PE) or polyvinylchloride (PVC). This allows for spans to be closer together (for polyphase lines) and greater sag with lower tension in conductors. Should they sway together, con-

Table 7-9. Characteristics of Conductor Materials (Commercial Grades)

Material	Conductivity, % (pure Cu = 100%)	Weight lb/in²	Weight lb/1000 ft per 1000 cmil	Ultimate strength, lb/in² (× 1000)	Elastic limit lb/in² (× 1000)	Modulus of elasticity (× 10⁶)	Temperature coefficient of linear expansion per degree (× 10⁻⁶) °C	Temperature coefficient of linear expansion per degree (× 10⁻⁶) °F
Copper—SD	99-100	0.320	3.027	36 to 40	18 to 20	12	17.1	9.5
MHD	98.5-99.5	0.320	3.027	42 to 60	23 to 33	14	17.1	9.5
HD	97-99	0.320	3.027	49 to 67	30 to 35	16	17.1	9.5
Aluminum, plain	61	0.0967	0.920	23 to 27	14 to 16	9	23.0	12.8
Aluminum, steel-reinforced	61	0.147	1.390	44	31	—	19.1	10.6
Steel	8.7	0.283	2.671	45 to 189	23 to 112	29	11.9	6.6
Copper-clad steel—30%	29.25	0.298	2.810	60 to 100	—	16 to 20	13.0	7.2
40%	39	0.298	2.810	60 to 100	—	16 to 20	13.0	7.2

To convert to metric system:

$lb/in^3 \times 0.0277 = kg/cm^3$

$lb/1000 \ ft \times 0.1488 = kg/km$

$lb/in^2 \times 0.0703 = kg/cm^2$

Courtesy The Anaconda Co., Wire and Cable Div.

Table 7-10. Characteristics of Solid and Stranded Conductors

Size	Cross section		Weight, lb/1000ft*		Resistance, Ω/1000 ft at 20°C		Solid conductor diameter, in	Stranded conductor	
	cmil	in²	Cu	Al	Cu	Al		Number and diameter of strands, in	Diameter, in
—	1,000,000	0.7854	3026.9	921.6	0.010	0.017	—	61 × 0.128	1.150
—	750,000	0.5891	2270.2	691.2	0.014	0.022	—	61 × 0.111	0.998
—	500,000	0.3927	1513.5	460.8	0.021	0.034	—	37 × 0.116	0.813
—	350,000	0.2749	1059.4	322.5	0.030	0.048	—	{37 × 0.097 / 19 × 0.136}	0.681 / 0.678
—	250,000	0.1964	756.7	230A	0.041	0.068	—	{37 × 0.082 / 19 × 0.115}	0.575 / 0.573
4/0	211,600	0.1662	640.5	195.0	0.049	0.080	0.4600	{19 × 0.106 / 7 × 0.174}	0.528 / 0.522
3/0	167,772	0.1318	507.9	153.6	0.063	0.102	0.4096	{19 × 0.094 / 7 × 0.155}	0.470 / 0.464
2/0	133,079	0.1045	402.8	122.0	0.078	0.128	0.3648	{19 × 0.084 / 7 × 0.138}	0.48 / 0.44
1/0	105,625	0.0830	319.5	97.0	0.098	0.161	0.3250	{19 × 0.075 / 7 × 0.123}	0.373 / 0.368
1	83,694	0.0657	253.3	76.9	0.124	0.203	0.2893	{19 × 0.066 / 7 × 0.109}	0.322 / 0.328
2	66,388	0.0521	200.9	61.0	0.156	0.256	0.2576	7 × 0.097	0.292
3	52,624	0.0413	159.3	48.4	0.197	0.323	0.2294	7 × 0.087	0.260
4	41,738	0.0328	126.4	38.4	0.249	0.408	0.2043	7 × 0.077	0.232
5	33,088	0.0260	100.2	30.4	0.313	0.514	0.1819	7 × 0.069	0.207
6	26,244	0.0206	79.5	24.1	0.395	0.648	0.1620	7 × 0.061	0.184
7	20,822	0.0164	63.0	19.1	0.498	0.817	0.1443	7 × 0.053	0.167
8	16,512	0.0130	50.0	15.2	0.628	1.030	0.1285	7 × 0.047	0.154

*For PE- and PVC-insulated conductors, add 550 lb per square inch of cross section for every 1000 ft.
To convert to metric system:
in² × 645 = mm²
in × 2.54 = cm

Courtesy The Anaconda Co., Wire & Cable Div.

tact will not result in conductors burning down.

Aluminum conductors steel reinforced are widely used for transmission lines; for long spans and high voltages, it is almost exclusively used. Such lines are subject to phenomena not usually experienced on shorter lines at lower voltages (including distribution circuits).

Skin Effect

In a conductor (of any material) carrying an alternating current, the self-inductance is more pronounced at the center of the conductor. Hence, the current flowing in the conductors will tend to flow more easily and a greater part near the surface of the conductor. This "skin effect" becomes even more pronounced at the higher voltages. For most conductors with comparatively small diameters, this effect is small and may be neglected. For transmission line conductors operating at high voltages and whose diameters are large, this skin effect becomes appreciable.

To accommodate this phenomenon, expanded conductors have been developed having hollow or partially hollow cores, eliminating the center part of the conductor not fully used in carrying current, Figure 7-21. While the greater overall diameter will expose the conductor to greater ice and wind loads, the advantage of the better current carrying ability and the lessened weight (compared to a non-hollow conductor of the same rating) partially compensates for the loading disadvantage.

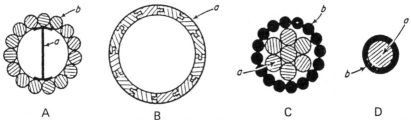

A   B   C   D

A   Hollow copper conductor with an I-beam core. *a*, I beam twisted spirally in a lay opposite to lay of individual wires *b*.

B   Type HH conductor using curved segments *a* laid in spiral configuration. Individual segments slide back and forth slightly when conductor is wound on reel.

C   Aluminum conductor steel reinforced commonly called acsr. *a*, steel strands making up the core; *b*, current-carrying aluminum strands.

D   One form of copperweld conductor. *a*, round steel core over which thin copper cylinder *b* is welded.

Figure 7-21. Conductors to Counter Skin Effect
(Courtesy Anaconda Co. -Wire & Cable Division)

Corona

In addition to the magnetic field about a conductor carrying an alternating current, there is also an electrostatic field. Such fields generally form in uniform patterns around a straight conductor and are also conductors of electricity. These patterns tend to concentrate and their conductivity ability at points where the conductor presents a sharp point, bend or comer. When the voltage exceeds a critical value, an energy discharge to the atmosphere takes place producing a luminous halo-like glow, known as corona, on the surface of the conductor. The diameter of the conductor, the conducting condition of the adjacent atmosphere, the condition of the conductor surface (such as roughness and dirt), and the presence of nearby conductors, all contribute to this effect. If the distance between the conductor and nearby conductors or structures is comparatively small, a sparkover may occur causing a short circuit and outage, and possible damage to the line. Corona discharges may be greater in rain, the drops clinging to the conductor change its shape encouraging corona where the drops act as sharp pips on the conductor. The hollow conductors of large diameter lessen not only the skin effect but that of corona as well.

As corona flashover may damage insulators to which conductors may be attached, particularly during rain, shields are provided at both the conductor and the supporting end of the insulator to furnish a path for the flashover away from the insulator, Figure 7-22.

One means of increasing the capacity of a transmission line, while at the same time minimizing corona losses is to replace the conductor with a larger one, or to add other conductors, in a "bundle" of two or more conductors on each phase held in place by spacers and suspended from each other a suitable distance, up to about 18 inches. Lower cost conductors may be used with this type construction, although greater ice and wind loads may be experienced and greater sag for a given space may result. Figure 7-25.

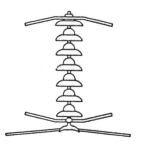

Figure 7-22. Arcing Rings or Horns to Keep Flash Away from Insulators. *(From Overhead Systems Reference Book)*

Galloping or Dancing Conductors

Conductors of overhead transmission lines where span lengths are relatively long and exposed in open country are subject to vibration and movement produced by wind.

One effect, known as aeolian vibrations, caused by wind eddies behind a conductor produces a regular high frequency oscillation of the conductor, whose frequency depends on the size of the conductor and the velocity of the wind. Masses of metal, known as dampers, are installed on the conductors at estimated node points to lessen the effects of these vibrations, Figure 7-23. The node points are impossible to determine precisely because the factors producing the vibrations are many and varying. The dampers, however, are placed near the towers within reach at points estimated to produce as much damping as practical. Armor rods are installed on aluminum conductors at the insulator clamps to reduce the wearing effect of the vibrations on the conductor.

To reduce the effects of aeolian vibrations, a self-damping type of conductor has been developed, Figure 7-24. By making the shapes of the conductor strands different, the outer strands trapezoidal while the inner ones are round, and relatively large clearances between them, the motion of the different strands tend to break up the vibrations. Appropriate splicing and terminating connectors and terminals are necessary.

A much more severe type of vibration, known as "galloping" or "dancing" conductors, is known to be caused by wind, but the mechanics are not always clear. One theory proposed that ice forming on the conductor approximates an airfoil and the wind blowing on it causes it to be lifted appreciably until a point where the conductor falls abruptly from the weight imposed on it, or is blown downward by the wind. The

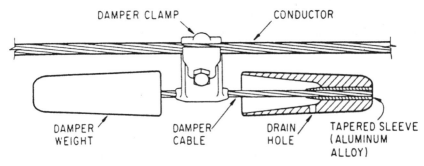

Figure 7-23. Dampers on a Conductor
(Courtesy A. B. Chance Co.)

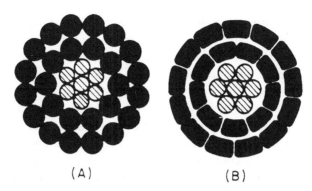

Figure 7-24. (a) Conventional ACSR; and (b) the Self-Damping Conductor. *(Courtesy Aluminum Co. of America)*

changing sag from the ice load coupled with the erratic nonrhythmic swaying may cause flashovers from conductors whipping together that may result in the burndown of conductors. The galloping of the conductors also causes extreme stress on both the conductors and supporting structures that may cause them to fail.

Although rare, unfortunately little can be done to remedy this condition except to attempt to melt the ice from the conductors. Resort is had to overloading the conductors temporarily by means of a "phantom" load connected to the circuit at the receiving end, or by transferring to it loads from other circuits, causing the conductors to become hot. Even the melting of ice on a portion of a span may cause the galloping to be interrupted. When possible, the circuit is taken out of service, allowing conductors to whip together without danger of damage or destruction.

Elastic Limit

The maximum stress to which conductors should be subjected is known as its elastic limit. It is a point at which the conductor can be stressed without permanent deformation, that is, the conductor returns to its original condition after being stressed. It is a point at which the conductor begins to elongate rapidly to failure. Elongation, even in a small percentage, results in comparatively large increase in sag. Sag is inversely proportional to tension, so that as the sag increases, the stress or tension in the conductor decreases. Hence, elongation is taken into account in determining allowable sag. For copper and aluminum, the elastic limit is reached at fairly low stress. The elastic limit of copper is generally from 50 to 60 percent of its ultimate breaking strength for medium hard drawn and hard drawn copper; for soft drawn, the elastic limit is very indefinite, but for practical purposes, is taken at about 50 percent of ultimate. For aluminum, the elastic limit is also indefinite, but is taken at 50 to 60 percent of the ultimate strength.

For ACSR, the elastic limit is a combination of the two metals. On initial stress, the two components will elongate equally and so will divide the stress in proportion to their cross sectional areas and their individual stress-strain characteristics. When a certain stress is reached, the aluminum portion will exceed its elastic limit and will elongate more rapidly in proportion to the increase in stress. The steel, having a higher elastic limit, will assume an increasingly proportion of the load until its elastic limit is reached. The elastic limit of the aluminum strands determine that for the complete conductor. If the steel strands are stressed beyond this point, the aluminum strands will loosen somewhat because of the increased length and will be on the road to failure.

Modulus of Elasticity

The modulus of elasticity is a measure of the way a conductor will sag under loading. A low modulus indicates a relatively large sag when the conductor is loaded, a high modulus indicates a comparatively small increase in sag between the highly loaded and fully loaded condition. Aluminum has a much lower modulus than copper and will show a greater sag when loaded with ice and subjected to wind. As line designs are usually based on ground clearances and 60° temperature, this factor should be taken into account.

Temperature Coefficient of Expansion

This characteristic is a measure of the change in length of a conductor with temperature. It is of great importance in determining the sag of a conductor at temperatures other than that in which it is strung. A high coefficient means a relatively large increase in length and sag; a high summer temperature may result in a greater sag of a conductor than under heavy ice loading. The coefficient of expansion of aluminum is greater than that of copper.

Connectors and Splices

Mechanical connectors and splices are almost universally used in connecting together of conductors. These not only insure good electrical conductivity, but also a uniformity in workmanship and mechanical strength. Some types are shown in Figure 7-25. These may consist of sleeves or yokes which are bolted, holding conductors together. Compression type connectors have the conductors inserted in a sleeve and the sleeve crimped by hydraulically made indentations. For ACSR conductors, two sleeves are used, an inner steel sleeve fitting over the steel core only, and an outer aluminum sleeve fitting over the entire conductor. In some instances, for ACSR conductors, only one outer sleeve is used and the indentations grip both the steel and aluminum conductors.

The indentations on a connector are often filled with solder and the whole splice polished to reduce the tendency for corona to form.

JOINT CONSTRUCTION

The use of a common pole for both power and communication lines is often done as a matter of economy and for better appearance.

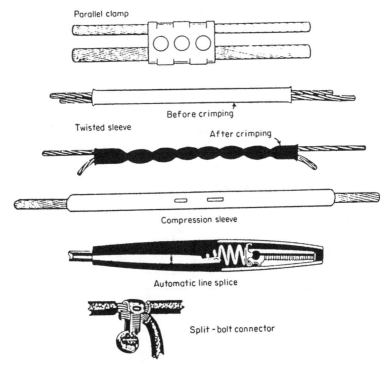

Parallel clamp

Before crimping

Twisted sleeve

After crimping

Compression sleeve

Automatic line splice

Split-bolt connector

Figure 7-25. Mechanical Connectors
(Courtesy Bundy Corp.)

The greatest use of this type construction is on the distribution system where poles may be shared with other users, telephone (the largest), telegraph, cable TV, traffic and lighting controls, fire and police alarms, etc. In rural areas where appearance may not be important, where long spans and lower clearances are permitted, and where services are relatively few, joint construction may not prove practical.

Generally, this type of construction may be desirable in those areas where facilities are located in streets and alleys, or on rear lot lines and easements from which services to consumers are extended. Such construction may result in heavier loading on poles, use of higher poles, a greater grade of construction, additional guying, more complex maintenance and coordination procedures between the users. Clearances between facilities and grades of construction, often greater than those recommended in the NESC are considered in the determination of space allotment and distribution of costs and savings.

The stresses imposed on the pole from all of the users must be taken into consideration. Some typical wind loadings for several telephone cables, with half-inch ice covering and a wind loading of four pounds per square foot are shown in Table 7-11. Tension acts on the pole from the messenger only.

Table 7-11. Loadings on Telephone Cables

Telephone cable sizes	Gauge (Cu)	Wind loading, lb/ft
50-pair including messenger	24	1.02
	22	1.03
200-pair, including messenger	24	1.23
	22	1.27
600-pair, including messenger	24	1.53
	22	1.54

Courtesy Long Island Lighting Co.

The space allocated for each purpose should be carefully defined. Communication circuits usually take up the lowest place on a pole and their sags taken into account in determining minimum ground clearance at the center of the span. Other communication circuits follow above with a "neutral" zone between these and the power circuit or circuits at the upper part of the pole. Although the neutral space is usually greater than that called for in the NESC for circuits of the voltages involved, a minimum of 40 inches is generally specified by telephone designers.

The division of space on a pole is usually based on the needs of each user of the pole. One method employs a "standard pole" in which the division of space is detailed, Figure 7-26. Since ground clearance is not the same for all locations, more space may be allotted to communication circuits than may be required. If more space is required by the power conductors, a higher pole may be required unless agreement can be reached lowering the communication allotment.

The division of costs may be more complex. Two methods are commonly used: the pole may be jointly owned, each user owning a share of the pole; or, the pole owned entirely by one user and space rented to the others. Such divisions appear equitable and easily accepted, but other factors serve to complicate the process.

In practice, power and telephone companies set, inspect and replace poles. Often, only the power company has the equipment to handle poles longer than 40 feet and is the only utility stocking them. In cases of emergency, the power utility must respond as quickly as possible while the other users can defer their work until after the power company completes its work. Poles on which power circuits operate at 5 kV or greater are handled by the power company; in replacement of such poles, the power company often cuts the old pole below the lowest

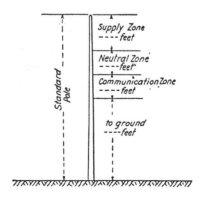

Figure 7-25. Allotment of Pole Space, on Standard Pole
(From Overhead Systems Reference Book)

power facility, allowing other users to relocate their facilities to the new pole, removing the stub of the old pole without coming near the power lines. Also, tree trimming costs are almost always assumed by the power company, although all users benefit.

Power lines at the top of the pole provide lightning protection for all the facilities farther down. Usually, one of the two major utilities obtains rights-of-way, permits, franchises, etc. On the other hand, power lines may cause interference with the operation of other users and may cause dangerous and widespread damage should they fall on the facilities of others. In event of injury or damage to workers and the public, each user must determine its financial liability. It is evident that agreements between the users for fair and equitable division acceptable to all is difficult of solution.

Although communication circuits on separate poles sometimes parallel transmission lines, they are seldom found on the same structures. They are sometimes installed on lower voltage transmission lines, but are sufficiently spaced from the power lines to avoid interference from the high voltage magnetic fields.

It is difficult, if not impossible, to deter the determined saboteur from plying his trade, particularly on the miles of exposed transmission lines. Some things can be done to make his task more difficult, to slow him down, and perhaps even catch him in the act. High fences with barbed wire tops and bottoms, Figure 4-18, no gates, access by bucket vehicles only, accompanied by battery operated sensors of intrusion, radioing back to the system operator and law enforcement agencies, and even electrifying the fences surrounding the bases of the structures supporting the lines may be more psychological than real deterrents, but might be worth employment. Other schemes and devices may come to mind. But the transmission lines are vulnerable from a distance. The solution appears to lie in the limiting of damage and the rapid restoration of facilities. Here, preplanned procedures, kept up to date should speed restoration.

Chapter 8

Underground Mechanical Design and Construction

Underground systems generally fall into two categories: facilities buried directly in the ground and those installed in ducts, manholes and vaults. Direct burial usually is employed in urban and suburban residential areas, while duct systems are confined to areas of high load density and areas where safety and environmental requirements make it desirable, if not essential. Application of these types of construction pertain to both transmission and distribution systems.

UNDERGROUND RESIDENTIAL DISTRIBUTION (URD)

The success of this type system is based on the development of plastics which are suitable for both conductor insulation and for the mechanical protection of cables. Coupled with improved plowing and trenching equipment and methods, this type construction is economically competitive with overhead systems.

Design

In areas employing this type of electrical supply, radial type distribution is usually specified. One pattern calls for a transformer to supply a number of consumers from secondary mains, Figure 8-1, while another calls for small transformers each supplying a single consumer, Figure 8-2. The first takes advantage of diversity between consumers' maximum demands, while the latter, usually requiring a greater total capacity of transformers to be installed, does away with secondary mains. A third pattern, somewhat of a compromise, calls for services only, two or more in number, to be supplied from one transformer without need of secondary mains.

Obviously, the number of consumers affected with the failure or de-energization of a transformer is different for each pattern. The choice of design will depend on safety, economy, service reliability and future requirements. In any of these designs, street lighting and other public parking, traffic lights, etc. are served much like other services.

It cannot be sufficiently emphasized that, while underground systems are less vulnerable to the vagaries of nature and humankind, when a fault develops on a cable under the ground, finding the fault and repairing it are more difficult and time consuming than similar faults on overhead systems. The possible exception occurs when a fault is readily apparent, such as one caused by a worker digging damaging a cable; but even here, repairs may be time consuming.

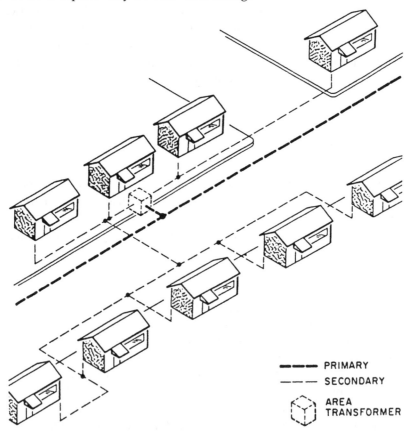

PRIMARY

SECONDARY

AREA
TRANSFORMER

Figure 8-1. Underground Residential Layout Using an Area Transformer *(Courtesy Long Island Lighting Co.)*

Like the overhead system, to maintain as high a degree of service reliability as practical, resort is had to duplicate supply, loop circuits, both open and closed. The primary circuits are therefore designed to provide tic points between circuits and sectionalizing facilities for isolating the faulted section and restoring service to the rest of the unfaulted circuit. While these essentially involve the same practices as are employed on overhead systems, the number of such ties and switching points is much greater, essentially one on each side of a distribution transformer. Moreover, the switching devices are more complex and expensive, and are not generally as easily accessible as those on overhead systems.

Where service reliability is of great importance in areas where such

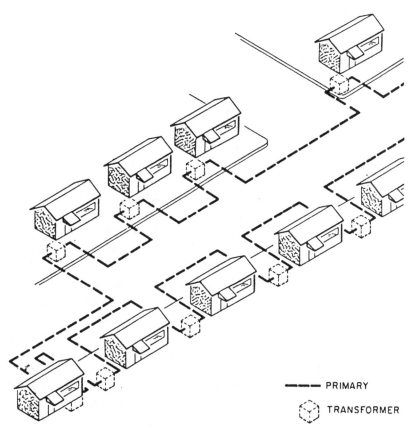

PRIMARY

TRANSFORMER

Figure 8-2. Underground Residential Layout Using Individual Transformers. *(Courtesy Long Island Lighting Co.)*

URD systems are installed (such as hospitals), duplicate supply generally is employed; networks, whether secondary or primary, are rarely considered.

Conductors

Conductors for URD systems are relatively simple in design usually consisting of a stranded conductor insulated by several forms of plastics, including cross-linked polyethylene (XLPE) and high molecular weight polyethylene (HMWP), sometimes also called high-density track resistant polyethylene (HDPE) for primary voltages and ethylene propylene rubber (EPR) for secondary voltages. These materials are generally used both as insulation and as protective sheathing. Earlier neoprene, polyvinyl chloride (PVC) and polyethylene (PE) sheaths gave way to the plastics mentioned, but many installations using these materials still exist.

Primary cables may have a layer of semiconducting material placed around the insulation to act as an electrostatic shield that tends to distribute electrostatic stresses in the insulation more uniformly. A plastic protective jacket is placed around the semiconducting layer to protect it from abrasion during handling and installation. Both primary and secondary multiconductors are each incorporated into single cables that, mechanically, are usually easier to handle, Figure 8-3a, b. A neutral conductor, which may consist of a bare circular or flat strap-shaped wire or ribbon is wrapped concentrically around the plastic cable and is an integral part of the cable. Electrically, the conductor and neutral arrangement results in reduced reactance and, hence, in a reduced voltage drop; the neutral also contributes to the electrostatic shield distributing such stresses in the insulation and reducing or eliminating this effect generally at bends that may cause insulation failure. The concentric neutral conductor also acts as mechanical protection during installation and for identification; round wires on secondary cables, flat shapes on primary cables.

Both primary and secondary cables are spliced in the same manner, Figure 8-3c. The conductors are connected together by an aluminum or copper sleeve which is then crimped. Insulation of the same material as that of the cable, preformed, of specified thickness, is placed around the connector; the insulation around the connector may also be built up by insulating tape of the same material to a specified thickness. The preformed insulation or the tape also acts as the protective sheath, except that in the case of primary cables, a semiconductor tape or preformed shape is placed between the insulation and the protective plastic sheath. The concentric neutrals are bundled together on one side and mechanically connected together and placed alongside the splice of the other conductors; they are then covered by plastic tape to protect them from corrosion and electrolytic action.

In some cases, the cables are connected to a completely insulated terminal block containing a number of stud connectors, Figure 8-4. The conductor is connected to one of the studs and the insulation and protective covering are taped to the insulation and molded covering of the terminal block. The terminal may also be made up of molded insulated load-break elbows and bushings. The conductor is connected to an insulated stud that fits into an insulated receptacle to make the connection. The conductor may be safely disconnected from the energized receptacle by means of an insulated hook stick. The molded terminal may also be

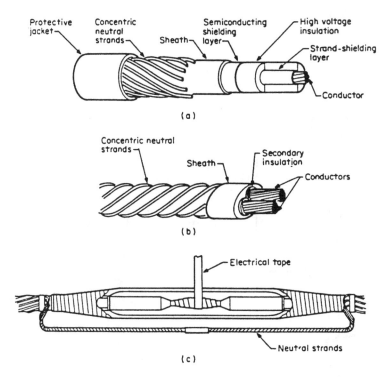

Figure 8-3. (a) Primary Concentric Neutral Underground Cable. (b) Secondary Concentric Twin Underground Cable. (c) Typical Single-Conductor-Cable Splice. *(Courtesy Long Island Lighting Co.)*

combined with similar type terminals on a transformer or switch. As mentioned earlier, they may serve as test points on loop circuits in restoring service after an interruption; they may also be used to rearrange, energize or de-energize primary circuits.

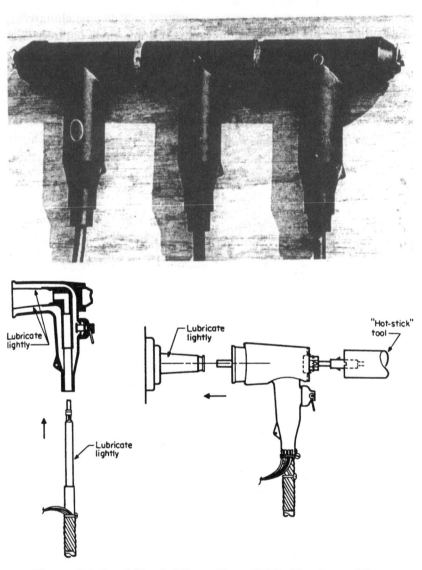

Figure 8-4. Load-Break Elbow-Type Cable Tap Assembly
(Courtesy Long island Lighting Co.)

Risers

Connection to overhead system with this type of cable is made by a clamp mounted on the end of the cabled conductor and the end taped with insulating tape of the same material used in the cable. Rain shields, shaped like cones, may be installed at the point where the conductor is attached to the clamp. The clamp may be operated from a "live-line" stick for connection to or removal from the energized overhead conductors, Figure 8-5.

Transformers

Transformers for the URD system are hermetically sealed against moisture including the terminals and bushings. The terminals may contain insulated disconnecting elbows providing a simple and flexible means of sectionalizing the primary circuit or disconnecting the transformer. See Figure 8-4 above.

The transformer may be installed on ground level pads of concrete, partially below ground in low profile semi-buried enclosures, or buried directly in the ground. Those on ground level or partially below ground sometimes have their connections made behind a protective panel with only the disconnecting elbow handles protruding so that no energized parts are exposed when the enclosure is opened; these are known as "dead-front" transformers. Figures 8-6 and 8-7.

Transformers may be of the so-called conventional type with associated fuse cutout or switches, or of the so-called "completely self-protected" (CSP) type described earlier. The units may have taps similar to overhead transformers. In rare cases, the underground transformers may

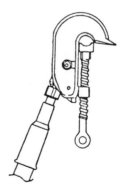

Figure 8-5. Live-line Or "Hot-fine" Clamp
(Courtesy A. B. Chance Co.)

require protection from lightning and surge arresters mounted on a pole or structure connected to the transformer terminals by way of a riser.

Transformer tanks and metal parts of the enclosures are usually connected to the system neutral conductor and serve as grounds for safety reasons. Often, separate connection to other ground is made to insure safety should the tank somehow become energized and the con-

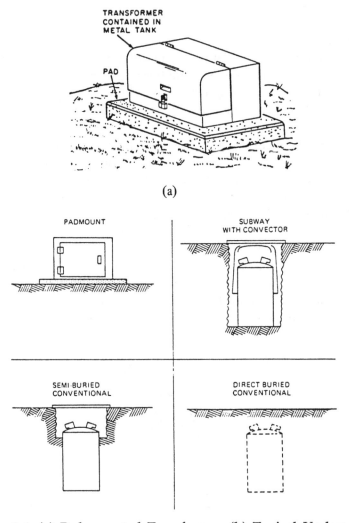

Figure 8-6. (a) Pad-mounted Transformer. (b) Typical Underground Installation. *(Courtesy Long Island Lighting Co.)*

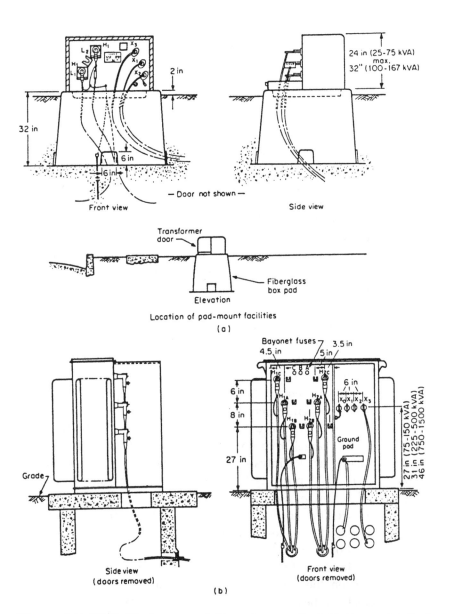

Figure 8-7. URD Underground Transformer Installations. (a) Pad-mount Transformer. (b) Three-phase, 4-kV Dead-front Metal-clad Transformer. *(Courtesy Long Island Lighting Co.)*

nection to the neutral become open or defective. In cases where the transformer is exposed to corrosion or may be so situated that a connection to the system neutral cannot be made, a separate bypass shunt is connected between the tank and the neutral.

Corrosion

Tanks of transformers, neutral conductors, and other metallic parts are subject to corrosion from chemicals and stray currents in the soil. This is especially true of equipment and cables buried in the ground, but also affects metallic items in duct and manhole type installations.

Transformer tanks may be painted or coated with plastic to inhibit corrosion from chemical and electrolytic action. Where these measures are ineffective, so-called sacrificial anodes connected to the tank and buried in the ground deliberately create an electrolytic action that causes currents to flow to the tank, rather than away from it, thus preventing a flow of metal ions away from the tank, the usual action of corrosion leading to the destruction of the metal. In doing so, the anode consumes itself and requires periodic inspection and replacement when necessary.

Stray currents may result from returns of direct current circuits (such as railroads), from galvanic action that takes place between dissimilar metals, especially in wet or moist soil or environment, from bacteria, or other causes. Two or more metallic objects, reasonably close together, immersed in chemical solution or vapor that may be present in soil or environment, will have an electric direct current flow between them. This current carries a flow of ions (molecules carrying an electric charge) from one to the other, a phenomenon known as galvanic action or electrolysis. Metal will then flow away from one metallic object and deposit itself on the adjacent metallic object. The direction of this flow of metallic ions depends on the relative voltages between the two (or more) objects, the flow being from the one of higher voltage to that of lower voltage. Typical voltages for different materials are shown in Table 8-1.

The deterioration of metal or corrosion that occurs from electrolytic action depends on: the direction, magnitude and duration of the current flow; the density of the current in the area in which the flow takes place; the moisture content of the earth or atmosphere with which the objects are in contact; and the chemical properties of the solutions or vapors through which the current flows.

Table 8-1. Galvanic Series

Material	Approximate potential with respect to a saturated Cu-$CuSO_4$ electrode, V*
Commercially pure magnesium	-1.75
Magnesium alloy (6% Al, 3% Zn, 0.1% Mn)	-1.6
Zinc	- 1.1
Aluminum alloy (5% Zn)	-1.0
Commercially pure aluminum	-0.8
Cadmium	-0.8
Mild steel (clean and shiny)	- 0.5 to - 0.8
Mild steel (rusted)	-0.2 to -0.5
Cast iron (not graphitized)	-0.5
Lead	-0.5
Tin	-0.5
Stainless steel, type 304 (active state)	-0.5
Copper, brass, bronze	-0.2
Mild steel (in concrete)	-0.2
Titanium	-0.2
High-silicon cast iron	-0.2
Nickel	+0.1 to -0.25
Monel	-0.15
Silver solder (40% Ag)	-0.1
Stainless steel, type 304 (passive state)	+0.1
Carbon, graphite, coke	+0.3

*These values are representative of the potentials normally observed in soils and waters which are neither markedly acid nor alkaline.
From *EEI Underground System Reference Book.*

The rate of penetration of a metal or metallic corrosion is roughly inversely proportional to the area from which the current discharge takes place, assuming the current is constant. The rate of corrosion, therefore, varies with the intensity of the current discharge. Theoretical rates of corrosion for some metals, in inches per year, are given in Table 8-2.

Table 8-2. Typical Rates of Metallic Corrosion

Anode metal	Density, lbs/in³	Penetration (in/yr) caused by discharge of 1 mA/in²
Magnesium	0.063	0.139
Zinc	0.258	0.091
Aluminum	0.098	0.065
Steel	0.284	0.071
Lead	0.409	0.182
Copper	0.323	0.142

From EEI Underground System Reference Book.

Where dissimilar metals are involved, a large anode should be connected to the metal of higher galvanic voltage (the emitting source) and a smaller cathode to the metal of lower galvanic voltage (the receiving metallic object). When insulating coatings are used to inhibit electrolytic action caused by dissimilar metals, preference is usually given to coating of the cathode rather than the anode.

The discussion on corrosion applies equally to URD systems where facilities may be buried directly in the ground and those where facilities are installed in ducts, manholes and vaults.

DUCT AND MANHOLE SYSTEMS

In areas of high load density where a multitude of large size conductors make overhead construction impractical, and in areas where the vulnerability of such facilities or where environmental reasons (trees, appearances, paved streets, etc.) also make such type systems undesirable, resort is made to underground distribution and transmission systems. Further, construction restraints, maintenance requirements, and, in some instances, economic considerations, make impractical the URD type of construction and operation. Here, cables are placed in ducts (or conduits), spliced in manholes and service boxes, and transformers and equipment installed in manholes and vaults.

The facilities installed depend not only on their function, but on the nature of the soil and terrain as well as subsoil obstructions, including the facilities of other utilities (gas, telephone, water, sewer, etc.). The ducts and manholes are generally made of reinforced concrete and may be prefabricated or constructed at the site, all in compliance with the NESC and other regional codes and regulations.

Ducts

Ducts are made of iron or steel pipe, fiber, tile, concrete, or other compounds, including plastics. Their usual diameters vary from 3 to 5 inches; some are encased in concrete, while others may be placed in well tamped sand or soil. They may be installed singly, but more often in duct banks of varying numbers and shapes determined by present and future requirements. The arrangement of the ducts in the bank may be affected by the space available and economics, but generally takes into consideration the dissipation of heat from the cables they enclose. Poor heat radiation, heat from too many cables in a duct bank, or from the character of the adjacent soil may actually limit the current carrying capacity of the cables to below their normal ratings. In general, the load carrying ability of cables, that is, the safe maximum operating temperature, will depend on their position in the duct bank. The relative advantages and disadvantages of some typical duct bank arrangements are indicated in Figure 8-8.

The depth to which ducts are to be placed should, if possible, be below the frost line to prevent dislocation from severe temperature changes. They should have a minimum depth to avoid possible damage from accidental "bull points." While following the natural or established grade of the street, they should be so installed that accumulated moisture drains toward the manholes.

The degree of curvature in duct banks should be kept as low as possible. In general, radii of curvature greater than about 300 feet and with ample clearance between the cable and duct, the cable pulling tension should not exceed acceptable limits, Figure 8-9. The bending radius depends on the maximum bending radius of the largest cable to be installed and should be from 7 to 20 times the radius of that cable, depending on its characteristics, that is, size, voltage classification, insulation, sheath, and other characteristics. The bends should preferably be located near the manholes as each bend will increase the pulling tension on the cable and thus reduce the maximum length or distance between man-

	Cost of duct construction	Ability to radiate heat	Cable support and racking conditions in manhole
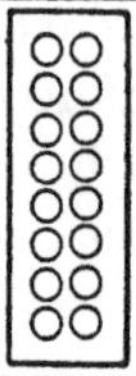	Expensive	Best	Best
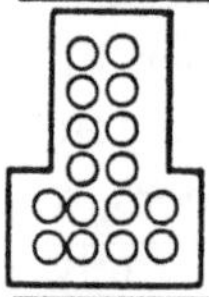	Moderate	Very good	Very good
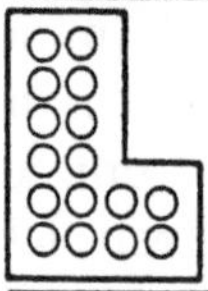	Moderate	Very good	Good
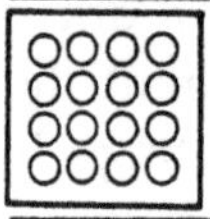	Cheapest	Very poor	Very poor

Figure 8-8. Comparative Duct Characteristics
(From EEI Underground Systems Reference Book)

holes. Reverse curves, especially as long duct runs, should be avoided wherever possible.

Where too many ducts enter a manhole, the congestion of cables may be intolerable and it may be desirable to build separate duct lines with separate manholes. For the safety of the worker, as well as for their efficiency, sufficient space must be provided for them.

Service Boxes

Ducts buried in relatively shallow depths terminate in service boxes and are usually located so as to accommodate secondary mains and the largest number of services without overcrowding the service box and without having too many bends in the service conduits. Usually constructed of precast reinforced concrete, they may be standardized in size, usually about four feet square and four feet deep. The entrance to them is usually quite large and square approximating the dimensions of the box, providing ample room for the worker to be able to stand erect

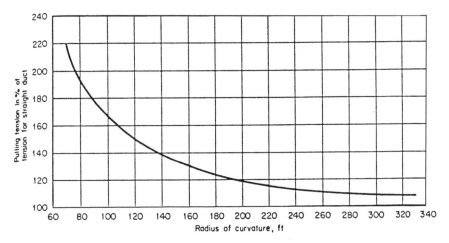

Figure 8-9. Effect of Radius of Curvature of Conduit on Pulling Tension (Applicable to a Conduit Consisting of One Continuous Curved Section). *(From EEI Underground Systems Reference Book)*

with the upper part of the body above ground, or to work from the ground level. Steel covers, with inner locked cover, keep out dirt and unauthorized persons.

Manholes

Larger than service boxes, cable manholes come in many sizes and shapes, some of which may be standardized, but generally shaped to accommodate the number and direction of the cables entering therein. Headroom of some six feet or more provides space for the worker to work safely and efficiently. They may accommodate secondary cables as well as primary and transmission cables; the latter proceeding from manhole to manhole bypassing service boxes, the ducts or conduits sometimes referred to as "trunks." Some typical shapes are shown in Figure 8-10. The various shapes take into account the training, splicing and racking of cables, the essential difference being the number of ducts entering the manhole and the angle at which they enter. The manholes are made of reinforced concrete, prefabricated or constructed in the field, and contain facilities for installing hangers to support cables and splices along the walls. The entrance, or throat or chimney, is generally wide enough to allow the entrance of workers and materials, and may be round or square, with similar shaped steel covers.

The manholes are spaced as far apart as required or as may be practical to hold down the number of cable splices. Location of the manholes should be such as to reduce to a minimum the number and radii of the bends in the duct system.

Generally, it is desired to keep the earth fill above the manhole at a minimum, not only for economic reasons, but to facilitate the installation of local services and to make more practical connections to street and traffic lights. Other subsurface structures and local regulations may sometimes dictate the actual depth at which the manhole roof may be located.

As water may accumulate at the bottom of manholes, some form of drainage needs to be provided. Where sewer connections exist, they are utilized. Where the bottom of the manhole may be below the natural water table, or where the earth may not support the manhole structure on the wall footings alone, means are provided to drain off water that may accumulate, in some cases by providing a dry well as part of the

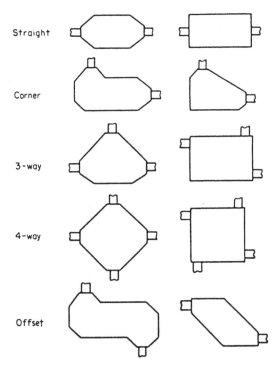

Figure 8-10. Typical Shapes of Cable Manholes
(From EEI Underground Systems Reference Book)

bottom of the manhole, or by means of a sump and pump automatically operated. Manholes are made as waterproof as practical, sometimes constructed of waterproof concrete, and painted with waterproof paint as added precaution.

Transformer Manholes

The dimensions of such manholes depend on the transformer and equipment they may contain, as well as their location. Like cable manholes, they may be standardized and prefabricated or constructed in the field. The dimensions should provide space, including sufficient headroom, for workers to be able to operate and maintain switches on both the primary and secondary facilities associated with the transformer. Features associated with cable manholes also generally apply to transformers; however, more often two openings or entrances to the manhole are provided, one at each end for ventilation in dissipating the heat from the transformer, with grates replacing the covers to the entrances of the manhole, Figure 8-11.

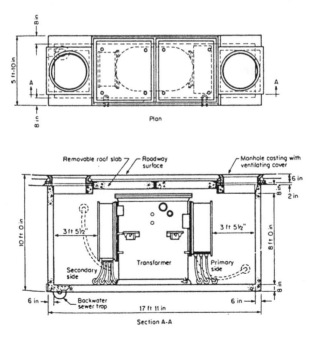

Figure 8-11. Transformer Manhole Under Roadway, with Removable Roof Slab. *(Adapted from EEI Underground Systems Reference Book)*

Manhole Design

The design of manholes follows design procedures usual for structures employing steel reinforced concrete. The methods described below incorporate data from several sources associated with typical manholes for electric systems. In designing a specific manhole at a specific location for a specific purpose, other local data, local rules and codes, local economics that include material and labor availability should be considered in any first design.

Reinforced Concrete Design

As the main parts of a manhole are made of reinforced concrete that reacts to stresses as rectangular beams, a review of the mechanics of beams may be useful.

The design of beams must insure against failure by compression, longitudinal tension and diagonal tension. Compression reinforcement is necessary where the dimensions of the members are limited. Longitudinal tension reinforcement is always required. Diagonal tension reinforcement, when necessary, may consist of steel stirrups, bent-up bars, or a combination of both.

Beam formulas are based on the assumption that concrete has no tensile resistance in flexure, that concrete is bonded completely to the reinforcing steel, and that no initial stresses exist.

Resisting Moments **(Refer to Figure 8-12)**

The resisting moment for tension, in terms of steel stress, is:

$$M_s = A_s f_s jd$$

and the resisting moment for compression, in terms of concrete stress is:

$$M_c = \frac{f_c kjbd^2}{2}$$

and for balanced design:

$$M_S = M_c = Rbd^2$$

where   As = area of longitudinal reinforcement
          fs = tensile unit stress in steel

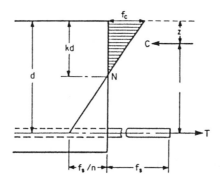

Figure 8-12. Distribution of Stress in Reinforced Concrete Beam
(From EEI Reference Book)

$\quad$ fc $\;=\;$ compression unit stress in concrete
$\quad$ jd $\;=\;$ arm of resisting couple
$\quad$ b $\;=\;$ width of beam
$\quad$ d $\;=\;$ effective depth of beam, from compression face to cen-
$\qquad\quad$ troid of steel

where $\quad R = f_s pj = \dfrac{f_c kj}{2}$

$\qquad k = \sqrt{2pn + (pn^2)} - pn = \dfrac{1}{1 + f_s/nf_c}$

and $\qquad j = 1 - \dfrac{k}{3}$

where $\quad$ p $\;=\;$ steel ratio A_s/bd
$\qquad\quad$ n $\;=\;$ ratio of modules of elasticity of steel to that of concrete

For fiber stresses:

$$f_s = \frac{M}{A_s jd} = \frac{M}{pjbd^2} \text{ and } f_c = \frac{2M}{jkbd^3}$$

For balanced reinforcement:

$$p = \frac{1}{\sqrt{(2f_s f_c)(f_s/nf_c + 1)}}$$

In a rectangular beam or slab, the shearing unit stress v is:

$$v = \frac{V_c}{\Sigma_0 jd}$$

and the bonding stress is:

$$u = \frac{V_c}{\Sigma_0 jd}$$

where V_c = the total vertical shear in concrete
        Σ_0 = the sum of perimeters of all longitudinal bars
        at a section

Web Reinforcement—Stirrups

Where the reinforcing bars are anchored at the ends, and the shearing stress exceeds $0.02f_c^1$ or $0.03\ f_c^1$ when the reinforcement spacing s is:

$$s = \frac{af_s jd}{V}$$

where   a = area of stirrup steel in one plane
        f_c^1 = compression unit stress in concrete at stirrup
        V^1 = total vertical stress at stirrup

When the stirrups are inclined 45°, s may be multiplied by $\sqrt{2}$ or 1.41.

Design Procedure for Rectangular Beams

1.   Determine the appropriate ratio b/d to be used. This may vary from 0.66 to 0.5 and may be less for longer beams. Determine the area bd required by the shearing stress v. Select b and d from above.

2.   From the equation for steel fiber stress f_s determine A_s and p. From the equation for steel ratio, for balanced reinforcement, determine the value of p. If the first value of p is equal or less than the second value of p, the tensile resisting moment governs and the depth d is satisfactory. If the first value of p exceeds the second value, increase d or provide compression reinforcement.

3.   Determine the bar sizes to provide the area of longitudinal reinforcement A_s and check that the width b will permit proper bar spacing. Bar spacing should be at least 2-1/2 diameters for round

bars; clear spacing between bars should be at least 1-1/2 times the maximum width of the aggregate particles used.

4.   Determine the shear and bond stresses to check that they are within the allowable values for the rectangular beam of the size selected.

Design Loading of Manhole
Live Load Criteria
The loading on the several parts of a manhole depends on the maximum load imposed on the street surface. The live load on the surface affects both the roof slab and the walls. Wheel loads of 21,000 lbs and impacts of 50 percent are standard for heavily traveled streets, and for conservative values, wheel areas of 6 by 12 inches or a surface area of 0.5 sq ft are considered. The concentrated load may then be:

$$\text{Concentrated load} = \frac{\text{wheel load} \times (1 + \%\text{impact} + 100)}{\text{wheel area}}$$

$$= \frac{21000(1 + 0.5)}{0.5} = 63000 \text{ lbs/sq ft}$$

The pavement, nature of the soil beneath, and the thickness (depth) of the soil above the roof of the manhole all serve to mitigate the actual effect of the concentrated load. The effective pressure is reduced at different depths below the surface, and these are shown in Figure 8-13 and Tables 8-3 and 8-4; these are based on a wheel load spread at a 45° angle for pavement and 30° for soil, the angles from the vertical in all directions. The wheel load area as a function of depth is for a 9-inch pavement.

Dead Loads
Typical unit weights for determining dead load are:
        Soil - 100 lbs/cubic ft wet soil weight
        Soil - 65 lbs/cubic ft submerged soil weight
        Pavement (plain concrete or asphalt)—144 lbs/cubic ft
        Reinforced concrete - 150 lbs/cubic ft

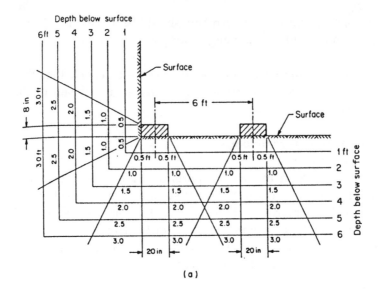

(a)

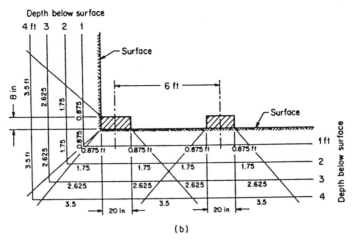

(b)

Figure 8-13. Diagram Showing Area of Spread of Wheel Loads (a) Based on 1:1 Spread and (b) Based on 1-3/4:1 Spread. *(From EEI Underground Systems Reference Book)*

Table 8-3. Pressure Calculations Based on a 21,000-lb Wheel Load—1:1 Wheel Spread

Depth from surface, ft	Area of spread		Area, ft²	Live-load pressure on cover*	Pressure on roof		
	Length L, ft	Width W, ft			Live load**	Surcharge	Total
0	0.67	1.67	1.12	18,500	5200	0	5200
1	1.67	2.67	4.46	4,710	2100	150	2250
2	2.67	3.67	9.80	2,140	1310	250	1560
3	3.67	4.67	17.10	1,230	960	350	1310
4	4.67	5.67	26.50	790	750	450	1200
5	5.67	6.67	37.80	560	620	550	1170
6	6.67	7.67	51.10	410	520	650	1170
7	7.67	8.67	66.50	320	450	750	1200

*Average pressure P_{av} that might be imposed on cover by maximum concentrated load, or (21,000 lb)/area.
**The surface concentrated load uniformly distributed over the width of the manhole, or $P_{av}W/6$.
From EEI Underground System Reference Book.

Table 8-4. Pressure Calculations Based on a 21,000-lb
Wheel Load—1-3/4:1 Wheel Spread

Depth from surface, ft	Area of spread		Area, ft²	Live-load pressure on cover*	Pressure on roof		
	Length L, ft	Width W, ft			Live load**	Surcharge	Total
0	0.67	1.67	1.12	18,500	5200	0	5200
1	2.42	3.42	8.27	2,500	1430	150	1580
2	4.17	5.17	21.5	970	840	250	1090
3	5.92	6.92	41.0	510	590	350	940
4	7.67	8.67	66.5	315	450	450	900

*Average pressure P_{av} that might be imposed on cover by maximum concentrated load, or (21,000 lb)/area.
**The surface concentrated load uniformly distributed over the width of the manhole, or $P_{av}W/6$.
From EEI Underground System Reference Book.

Allowable Stress Bases

1.   Allowable Stresses - Concrete

Type of Concrete	n	f_c^1	f_c	V_c
Precast plant	8.5	3500	1575	118
Precast plant	7.0	5000	2250	141
Field placed	8.5	3500	1100	85

2.   Reinforcing Steel
      For ASTM A615: grade 40 deformed-billet bars
      f_s = 20000 lb/sq in.; for grade 60 f_s = 24000 lb/sq in.

3.   Structural Steel
      All solid steel covers, gratings, and other structural elements sub-
jected to repeated traffic loading, design in accordance with American
Association of State Highway and Transportation Officials (AASHTO)
Requirements for Design of Repeated Loads for 500,000 cycles of load, and
with the latest revision of ATSC *Manual of Steel Construction.*

4.   Soil Bearing Pressure
      A conservative value of 1.5 tons/sq ft may be used unless organic
clays or silts are involved. If a manhole or vault is to be installed on clay,
claying soils, or organic material, careful evaluation should be made of
the potential for settlement. Use of crushed-stone base or piles may be
required and soil bearings may be necessary.

Wall Design

      Manhole wall designs are based on the longitudinal component of
the effect of both live and dead loads acting on the walls. The horizontal
forces will depend on the surface, the angle of repose of the soil, and the
effect of the water table.
      At depths below about 5 feet (as shown in Table 8-3 for the spread
of wheel loads), the weight of the earth above the manhole predomi-
nates. Here, the average of the live-load effects approximate 450 lb/sq ft
and appears to be constant for lower depths.
      The dead loads at the various depths and various horizontal pres-
sures as a percentage of the vertical pressure are shown in Table 8-5,
which extends the tabulation associated with Figure 8-13 and Table 8-5

also serves as a guide in determining the horizontal pressures with various headrooms and depths for the several corresponding angles of repose of the soil and pressures from the hydrostatic head of the water table.

Table 8-5. Horizontal Earth Pressures at Various Depths

		No live load			Live and dead loads				
		Horizontal pressure, lb/ft^2				Total	Horizontal pressure, lb/ft^2		
Depth	Dead load	25%	30%	35%	Live load 1-3/4:1	live and dead load	25%	30%	35%
0	0	—	—	—	5200	5200	—	—	—
1	150	38	45	53	1430	1580	395	474	553
2	250	63	75	88	840	1090	273	327	382
3	350	88	105	123	590	940	235	282	329
4	450	113	135	158	450	900	225	270	315
5	550	137	165	193	450	1000	250	300	350
6	650	162	195	228	450	1100	275	330	385
7	750	187	225	262	450	1200	300	360	420
8	850	212	255	298	450	1300	325	390	455
9	950	237	285	333	450	1400	350	420	490
10	1050	263	315	367	450	1500	375	450	525
11	1150	288	345	402	450	1600	400	480	560
12	1250	312	375	438	450	1700	425	510	595
13	1350	338	405	472	450	1800	450	540	630
14	1450	352	435	507	450	1900	475	570	665
15	1550	387	465	542	450	2000	500	600	700

From EEI Underground System Reference Book.

Rigid Horizontally Reinforced Frame

The wall loading for lateral earth pressures from live load and dead loads may be taken from the chart in Figure 8-14. The frame may be analyzed using conventional intermediate structural techniques. Midspan moments and corner moments may be calculated from the formulas noted earlier, or by using the coefficients for each moment given in Figures 8-15, 8-16, and 8-17.

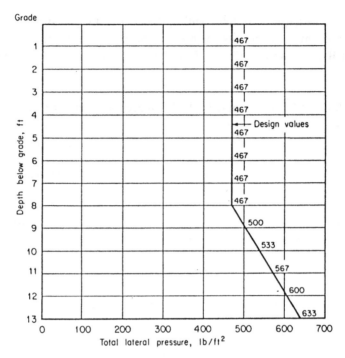

Figure 8-14. Design Chart-Lateral Pressures on Walls
(Courtesy Consolidated Edison Co.)

Simple Vertically Reinforced Structure

The wall loading for lateral earth pressures due to live and dead loads may be taken from the chart in Figure 8-14. The wall may be analyzed as a simply supported strip with a height equal to the headroom of the manhole plus one-half the sum of the floor and roof thickness. This method should be used with field-poured manholes *only* and requires a field-poured roof connected to the walls and able to carry the wall reaction.

Combination of Horizontally Rigid Frame and Vertically Reinforced Design

A combination of the two methods described above may be used when conditions indicate there are areas where the reinforcing for both the vertical and longitudinal methods is severely interrupted by openings.

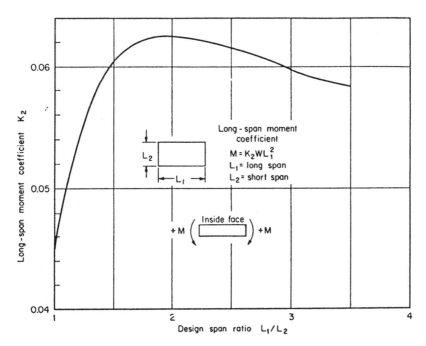

Figure 8-15. Long-Span Moment Coefficient
(Courtesy Consolidated Edison Co.)

Partition Walls in Vaults
Partition walls in field-poured vaults that house oil-type transformers inside consumer's property may be designed for an internal blast load of 600 lb/sq ft.

Other Requirements
All field-poured, vertically-reinforced manholes and vaults should have a minimum thickness of 6 inches; field-poured, horizontally-reinforced rigid frame type manholes and vaults should have a minimum thickness of 8 inches.

Where watertight construction is required, the walls may be monolithically poured with the flow a minimum distance of 12 inches above the top of the floor level. A water stop should be inserted at this location, as shown in Figure 8-18. A minimum wall of 10 inches is required for vault construction under those conditions, and 5000 lb/sq in. concrete may be used.

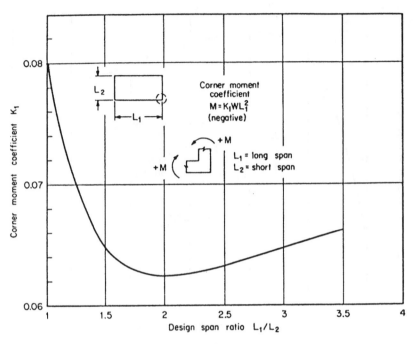

Figure 8-16. Corner Moment Coefficient
(Courtesy Consolidated Edison Co.)

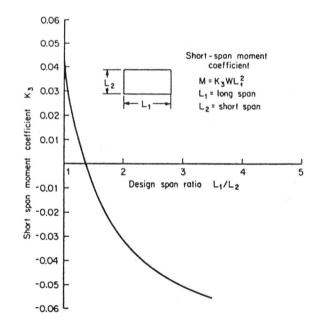

Figure 8-17. Short-Span Moment Coefficient
(Courtesy Consolidated Edison Co.)

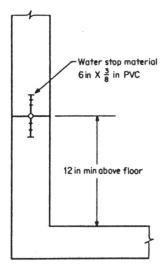

Water stop material
6 in X $\frac{3}{8}$ in PVC

12 in min above floor

Figure 8-18. Water Stop Detail; Water Stop to Be Lap-Spliced 4 in. on Each Side of Vertical Joints and Continuous Around All Corners. *(Courtesy Consolidated Edison Co.)*

Roof Design

Manhole roofs may be designed as a series of structural steel beams or rails, or reinforced concrete with extra-heavy steel reinforcement or structural steel to support the manhole frame. Where installed in sidewalks or other areas not subjected to heavy vehicular traffic, roof designs may take into account the lighter loading. If there is any question of loadings, the heavier loading design should be used.

Live Loads

Roof structures for manholes or vaults may be designed to carry the live loads specified above.

Wheel-Load Area

The wheel-load design may be taken acting on areas that may be determined using the method of spreading a concentrated load defined above under Wheel-Load Distribution.

Field-Poured Manholes

Field-poured manhole roofs may be designed using structural steel sections around the roof opening to support the manhole frame.

Precast Manholes

Precast manhole roofs may be designed using a simply supported reinforced concrete beam around the opening to support the manhole frame.

Roof Slabs

Roof slabs may be designed as one-way or two-way reinforced concrete slabs; where the ratio of short span to long span exceeds 0.5, a one-way slab design may be used.

One-way Slab—The design moment may be determined using a simply supported beam loaded with the effective live-load intensity and uniform dead load. Design moments as a function of depth are given in Table 8-6.

Table 8-6. Simple Support Roof Slab Moment, ft-lb
(From EEI Underground Systems Reference Book)

Design depth, ft	Design span				
	4ft	*5ft*	*6ft*	*7ft*	*8ft*
0.75 to 1.75	5,500	7,860	10,270	12,740	15,260
1.75 to 5.0	2,860	4,690	6,330	8,130	10,460

Two-way Slab—The design moment may be determined using Table 8-7 for a one-way stab and proportioning the one-way slab design moments for the short-direction and long-direction.

Table 8-7. Conversion Factors for Two-Way Slab Moments
(From EEI Underground Systems Reference Book)

Ratio of clear spans	Long-span moment factor K_1	Short-span moment factor K_3
1.0	0.500	0.500
0.9	0,396	0.604
0.8	0.295	0.709
0.7	0.194	0.806
0.6	0.114	0.886
0.5	0.059	0.940

Short-span moment $M_3 = K_3 \times$ simple-span moment
Long-span moment $M_1 = K_1 \times$ simple-span moment

Above Grade Vault Roofs

Above grade vault roofs may be designed for a uniform dead load plus a live load of 30 lb/sq in. of projected area plus internal blast load.

Other Requirements

The minimum thickness of a precast roof should be 6 inches, and 8 inches for a field-poured roof.

Floor Design

In the design of manhole floors, the load-bearing power of the soil and the height of the water table play an important part. The soil must support the weight of the manhole structure, its contents, and any imposed surface live loads. In firm soils, the earth is capable of supporting the structure and any additional weight. The floor, therefore, is often poured after the walls are in place, adding to the strength of the walls. Floor walls may be 4 to 6 inches thick.

Where the earth is not capable of supporting the loading of the walls, the floor is used as a means of spreading the load. Here, the floor is poured before the walls are installed. Similar measures are employed in areas of high water table. Such floors are usually made of reinforced concrete, a minimum thickness of 6 inches, and are constructed with a keyway for the walls. Where the hydrostatic pressure may be high, an additional pour of 2 to 4 inches of concrete is added on top of the floor.

Prefabricated manholes may be completely precast in one piece, or in a caisson type in which the roof and floor are separate. The caisson walls may be sunk in place, the precast floor may be placed within it (or a floor may be poured), and a precast roof installed in keys in the walls provided for that purpose (such roofs may also be installed in other types of manhole construction). Small manholes or service boxes may also be completely precast or formed from precast pieces.

Frames and Cover Design

Frames and covers may be made of cast iron, malleable iron, or steel, and designed to withstand the loadings described earlier, and covers infrequently used may be made of reinforced concrete. Depending on the area over which the load may be applied, frames and covers may have to withstand wheel loads from 50,000 to 200,000 pounds, although sidewalk covers may be designed for lowered loadings. Covers may be square or round, but the latter is preferred to insure against their falling into the manhole when being replaced. Frames and covers for transformer manholes may be of the completely prefabricated grating type, or of the combination type, a part solid and a part grating, that may be specified for roadway use, but may be of lower loading rating.

Transformer manholes are usually built with a removable roof slab covering an opening capable of admitting large distribution units and other equipment. The manholes are of reinforced concrete with slabs sealed and made watertight, the pavement being replaced after the transformer and equipment are installed. The pavement is removed and the slab removed when transformer replacements are required. When the transformer manhole is located under the sidewalk, the roof slabs are flush and made part of the sidewalk surface, and are readily removed when necessary. Prefabricated manholes may be completely precast or formed from precast individual walls, roofs and floors.

Transformer Vaults

Where transformers are to be installed inside a building, or some-times under a sidewalk, vaults are constructed. They are usually of re-inforced concrete, the dimensions of which depend on the transformer and equipment to be installed, the space available, and its location, the adjacent structures and substructures, and the applicable codes and local ordinance requirements. Access for both equipment and personnel from the outside is usually provided, but adequate internal access may be provided instead. In very tall buildings, such vaults may be located on upper floors as well as in the basement. The method of supply is some-times known as vertical distribution. Ventilation requirements are the same as for transformer manholes. Vault ceilings or roofs should be designed to take care of possible fire and explosion from transformer failure or other causes.

Ventilation

The main source of heat in a transformer manhole stems from the losses in the core and windings of the transformer, losses that can be calculated or included in manufacturers' specifications. The dissipation of heat is generally based on the area of the enclosing walls and the nature of the adjoining soil conditions. For proper operation of trans-formers, the manhole should have sufficient volume, or cubic content, supplemented with natural ventilation, to keep the transformer within prescribed temperature limits. Air temperature in the manhole should not exceed 40°C, generally occurring at periods of maximum load. The appropriate number of cubic feet per minute of air to dissipate the heat may be found in the curve of Figure 8-19. When such limits cannot be attained by normal circulation of air from the two gratings of the transformer manhole, it may be necessary to provide some means of forced ventilation, usually in the form of fans or blowers. The same ventilation and ventvolume requirements apply to transformer vaults as for transformer manholes.

Summary

Design drawings and stress diagrams for a typical manhole are described and shown in Figure 8-20.

DISTRIBUTION CABLES

Conductors for use in underground systems must be provided not only with insulation sufficient to withstand the voltages at which they operate, but with some kind of protective sheath; this combination or assembly is generally referred to as a cable. Practically all cables consist of a copper or aluminum conductor surrounded by a plastic material which serves for both insulation and protective sheath, a type of cable used for both primary and secondary circuits and was described earlier in connection with URD systems. Figures 8-21 a and b.

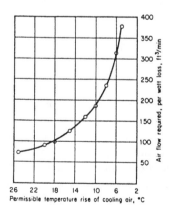

Figure 8-19. Airflow Requirements for Limiting Temperature Rise in Transformer Vaults. *(From EEI Underground Systems Reference Book)*

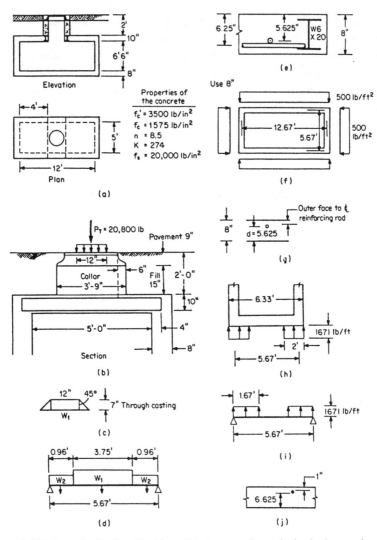

Figure 8-20. Sample Design Problem. (a) Assumed manhole design and properties of the concrete. (b) Assumed manhole roof design and loading. (c) Diagram showing assumed load spread through casting at top of manhole collar. (d) Design showing dead loads imposed on manhole roof. (e) Assumed roof slab design showing steel beam and reinforcing rods. (f) Approximate wall dimensions and assumed lateral pressures on walls. (g) Assumed reinforcing rod cover for 8-in. concrete beam. (h) Assumed manhole floor design and reaction areas. (i) Diagram showing moments acting on manhole floor. (j) Assumed reinforcing requirements for 8-in. floor slab. (*Courtesy Consolidated Edison Co.*)

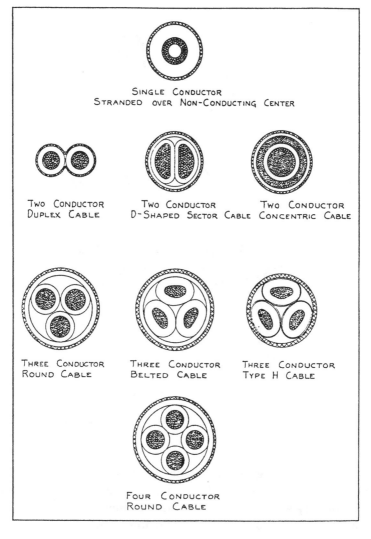

Figure 8-21 a. Cross Section of Typical Cables
(From EEI Underground Systems Reference Book)

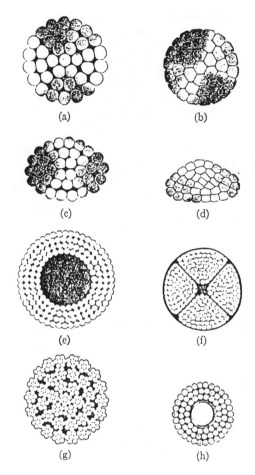

Figure 8-21b. Cable Conductors

(a) Standard concentric stranded   (e) Annular stranded (rope core)
(b) Compact round   (f) Segmental
(c) Non-compact sector   (g) Rope stranded
(d) Compact sector   (h) Hollow core

Photographs in this figure furnished by the Okonite-Callender Cable Company *(From EEI Underground Systems Reference Book)*

For many years prior to the introduction of the plastic insulated and sheathed type cable, insulation of rubber, varnished cambric, and oil impregnated paper were used in cables with a sheath of lead. Many of such cable installations exist and will continue to exist for many years to come. In some circumstances, cables of this type may be replaced with similar cables.

Rubber insulation is almost exclusively confined to consumer service and secondary mains of voltages of 600 volts or less, and to some primary cable applications of voltages of 5 kV and below. Because rubber insulation sometimes contains sulfur compounds (to extend its usefulness) that react destructively with copper, the strands are often tin-plated, adding to the complexity and cost of rubber-insulated cables.

Varnished cambric is used as insulation for primary and subtransmission cables for voltages up to about 45 kV. Introduced when primary distribution voltages of the 15 kV class were adopted, this insulation gave way to oil-impregnated paper cables which were not only adaptable to higher voltages, but easier to handle and less costly.

Single conductor cables are more flexible and generally easy to handle than multiconductor cables. Hence, they are used for secondary mains and services, for primary line branches (spurs or laterals), street lighting, and other purposes where many branch joints or splices are required, or when the size of conductor is so large as to make multiconductor cable impractical.

Multiconductor cables of two, three, and four conductors are used in the main portion or trunk of primary feeders where relatively few branch connections are required, and more generally used for two- and three-wire services. Multiconductor cables are more economical from material and labor standpoints than a number of single conductor cables.

In some instances, lead sheathed cables may be buried directly in the ground or placed under water in submarine installations. Here, the cable may be covered with jute or tar and armor wires of steel wound around the whole to protect the cable from mechanical injury.

When the cables are installed in ducts or conduits, a sufficient clearance must be provided between the walls of the duct and the cable or cables. The minimum clearance depends on the length of the duct run, the diameter of the duct, the number and curvature of duct bends, and the quality and alignment of the duct sections. In general, the duct diameter should be at least a half-inch larger than the cable, or the (imaginary) circle enclosing several cables where such are installed. In some cases (such as a single cable installed in a straight, clear, duct), a lower clearance may be acceptable provided the pulling stresses do not approach values that will damage the cable. A single cable in a duct requires less tension, weight for weight, than several cables in a duct, particularly when they are of large size and heavy weight.

In some special instances, cables may be placed in troughs located under sidewalks. Sidewalk slabs act as covers for the precast reinforced concrete troughs.

Joints or Splices

Sections of cables installed in underground ducts are connected in the manhole to form a continuous length; the connection is called a joint or splice. It is possible to connect more than two cables together, and these are referred to as three-way, four-way, etc. splices.

Joints or splices in cables with nonplastic insulation and metallic (usually lead) sheaths are more complex and require more time and greater skills to make them than the plastic types described earlier. The conductor connection is a tube usually made of copper or aluminum matching the conductor material. The tube is crimped on to the ends of the two conductors to be joined. Some of the older connectors used split sleeves with the conductors squeezed into the ends and the horizontal split filled with solder, covering the conductors; this type is obsolete, but many still exist. A lead cylinder is slipped over the connections after they are made and the two ends of the cylinder are wiped to the sheaths of the two cables being spliced. Two holes in the sleeve allow molten insulating compound to be poured into the assembly, one hole allows air to escape. The whole completed joint may be covered with a fireproofing material, usually a mixture of sand and cement. This is known as arc-proofing, to prevent the spread of fire or explosion should a splice or cable fail. A drain or ground wire is sometimes wound around the joint before arc-proofing and connected to an electric ground. A typical splice of this type is shown in Figure 8-22.

As the cable expands and contracts with cycles of load, the sheath may crack and connectors loosen. So that this movement may take place safely, the cables are racked along the walls of the manhole. A large reverse curve is made in the cable passing through the manhole. The large radius 90° bends enable the cable to take up the expansion movements taking the stresses off of the splice. Splices are supported on racks

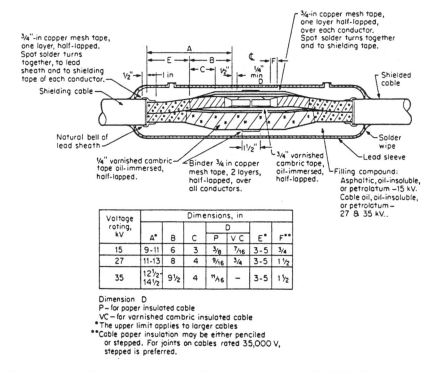

Figure 8-22. Straight Joint for Three-Conductor Shielded Paper or Varnished-Cambric-Insulated Lead-Covered Cable, 15 to 35 kV. *(From EEI Underground systems Reference Book)*

mounted on the walls again taking stresses from the splice. Cables and joints mounted on the walls prevent damage from materials that may fall into the manhole when covers are removed. Finally, the cable at the splice should bear some identifying mark, usually a tag indicating the feeder designation, size, voltage, etc., tied to the splice.

Underground to Overhead Connection

In many instances, underground cables are connected to overhead lines through a riser. This consists of leading the cable through a curved length of pipe fastened to the side of the pole. For plastic-insulated cables, the end of the cable conductors are wrapped in plastic insulation tape and the conductors terminated in clamps that fasten on to the overhead conductors; this was described earlier in connection with URD systems.

When the insulation of cables is not a plastic, as in older installations, the cable is terminated in potheads for primary cables and weather-heads for secondary cables, Figure 8-23.

Cable sheaths are attached to potheads by clamping devices, or by wiped joints. The conductor or conductors of the cable are connected to terminals inside the pothead which are brought outside of the pothead through bushing type terminals. The pothead body is filled with a liquid insulating compound that cools into a solid. The overhead wires are connected to the female end of the terminal, enclosed in an insulating cap. This type pothead is known as a disconnecting pothead to distinguish it from an ordinary pothead where the connections are made di-

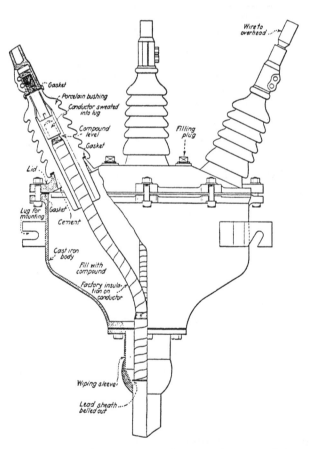

Figure 8-23a. Makeup of One Type of Vertical Pothead for 4,800 Volt or 6,000 Volt Service. (*Courtesy G. & W. Electric Specialty Co.*)

rectly to the terminals extending from the pothead case.

A simpler device is used in the case of secondary cables of 500 volts or less rating. The underground cable conduits are brought out through a preformed insulator, usually of porcelain, in an assembly that inverts the leads to prevent rain entering the cable or riser. This device is known as a weatherhead and connection to the overhead conductors is made with ordinary connectors of several types described earlier.

TRANSMISSION CABLES

Underground transmission cables may be buried directly in the ground, may be installed in ducts, or may be contained in pipes buried

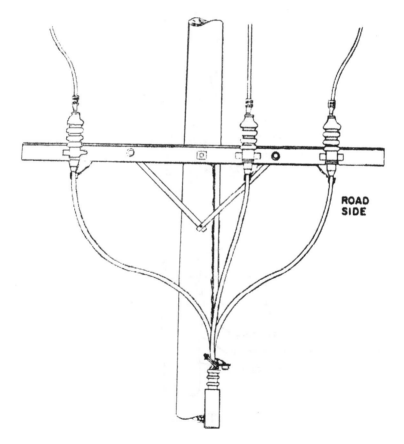

ROAD
SIDE

Figure 8-23b. Outdoor Terminals (Potheads) for 13 kV Cables.

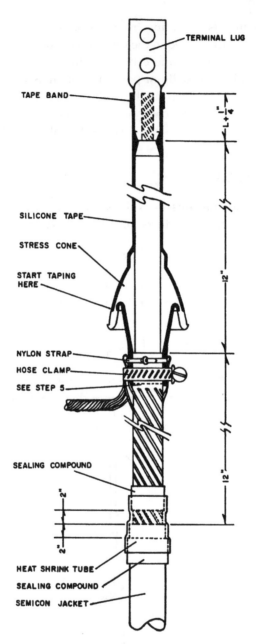

TERMINAL LUG

TAPE BAND

$L + 1\frac{1}{4}"$

SILICONE TAPE

STRESS CONE

START TAPING
HERE

12"

NYLON STRAP

HOSE CLAMP

SEE STEP 5

SEALING COMPOUND

12"

2"

2"

HEAT SHRINK TUBE

SEALING COMPOUND

SEMICON JACKET

Figure 8-23c. Indoor Terminal for 13 kV Cable.

in the ground. The kind of installation depends on the voltage and type of cable as well as the area in which it is located.

Transmission cables may range in voltage from 33 kV to 500 kV, the lower voltages generally applying to subtransmission circuits. Insulation for cable rated to about 138 kV may be of the solid type, using oil-impregnated paper insulation on older types to 46 kV and cross-linked polyethylene to 138 kV on newer installations. Above the 138 kV level, cable insulation consists generally of oil-impregnated paper under oil or gas pressure in so-called hollow-type cables; in older installations, the hollow-type included cables rated from 69 kV and higher.

Generally, cables of 33 kV and 45 kV rating, containing three conductors and lead sheathing, were constructed with an outside diameter that permitted their installation in a duct. Cables rated to 138 kV, single conductor, of the hollow type, could also be installed in ducts. Above these ratings, three single conductor, oil-impregnated paper insulated cables, are installed in pipes of larger diameter, with usually longer sections between manholes. All of these, however, may be buried directly in the ground, with steel armor or some kind of protection on the cable or pipe.

Transmission cable insulations present particular problems because of the electrostatic stresses imposed on them by the high operating voltages. Because of the mechanical stresses that may be created not only from the installation process, but also from the expansion and contraction of the cable (and insulation) caused by the cycles of load, minute voids are formed in the insulation. This is further aggravated by the skin-effect of the conductor in which the greater part of the current flowing tends to flow near its surface producing heat that may have additional destructive effect. Under the high voltage, the voids or air pockets tend to ionize and become conductors of electricity; the associated corona effect carbonizes the insulation, resulting in tracking or progressive breakdown of the insulation and ultimate failure of the cable. To overcome this destructive effect, the higher voltage cables maintain the insulation under insulating oil or gas pressure that fill the voids, preventing ionization from occurring and restoring the insulation to its former high value. The gas employed is usually nitrogen, but may be sulfur hexafluoride.

The oil or gas pressure may be applied to the cable insulation in one of two ways. The first (and original) method, the cable has a hollow core formed by the conductor strands wrapped around a helical ribbon, the hollow core containing the oil or gas under pressure. In the second method, the cable with solid insulation is installed in a pipe filled with oil or gas under pressure, Figure 8-24a, b, c. Where parallel circuits exist, the oil or gas may be circulated between the two circuits and through heat exchangers situated at each end, cooling the conductors and increasing their current carrying capability.

The installation and maintenance costs of the hollow core type cable may be greater than those of the pipe type cable, but the hollow

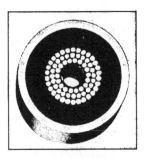

(a) A 120-kV single-conductor oil-filled cable has a stranded-copper conductor surrounding a spiral-steel open core. Oil passes through the open core to permeate the cable at 10 to 20 psig pressure. *(Courtesy Pirelli Cable Corp.)*

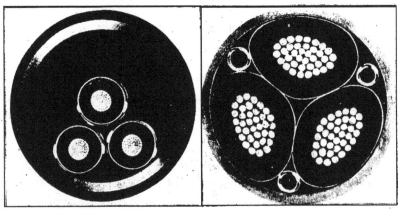

(b) Section through high-pressure gas-filled pipe cable for 115 *M* Each conductor has skid wire wrapped around it. Nitrogen at 200 psig fills the pipe interior. *(Courtesy General Cable Company)*

(c) Three-conductor gas-filled cable for 69 *M (Courtesy General Electric Co.)*

Figure 8-24.

type cable takes advantage of the skin effect in the conductors carrying alternating current, particularly at high voltages. In addition, the cooling effect of the oil or gas in contact with the conductors in hollow type cables, compared to that of the pipe type cable in which the oil or gas is in contact with the insulation, further increases its current carrying capacity. Because of the type of construction, the hollow core cable may be installed in a duct of a duct bank, whereas the pipe type cable requires the construction of a separate pipe line. Both types may be buried directly in the ground; the hollow core cable may require armor and waterproof covering while the pipe has a fibrous covering to prevent corrosion and electrolytic action. Care should be taken when pulling the cable into ducts, pipes or trenches not to impose stresses that may damage the cable; dynamometers should be used to ascertain that allowable pulling stresses are not exceeded.

Since the soil in which the cables or pipes may be buried may not provide adequate heat dissipation, soil adequate for this purpose may need to be placed surrounding the cable or pipe where hot spots are likely, in order not to affect the current carrying capacity of the cable. There are thermal sands available for this purpose.

Cables of this type have been used in circuits operating at 69 kV to 500 kV. Special accessories and auxiliary equipment to handle the oil or gas is necessary with such cables.

Joints

Cables of both the hollow core and pipe type may be spliced in a manner similar to those made on solid type cables. Where such installations are long, an oil or gas leak could result in long and costly repairs if it became necessary to decontaminate the oil or gas of the entire length of the circuit. To obviate this possibility, the length of the circuit may be sectionalized by means of joints designed for that purpose called stop joints and semi-stop joints. The number and type of such joints depends in large part on the length and importance of the circuit.

Stop joints sectionalize both the conductors and the oil or gas flow by means of a physical barrier. The sectionalizing enables both the conductor and oil or gas systems to be repaired without affecting the entire line. The semi-stop joint does not sectionalize the conductor of the cable but provides a physical barrier to the flow of oil or gas. Typical joints are shown in Figure 8-25a, b, c, and d.

In the repair of any part of such cables, only the oil or gas in the

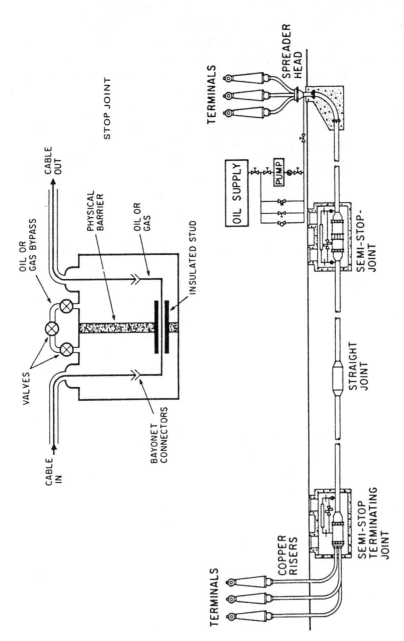

Figure 8-25a. Simplified Diagram Showing Methods of Isolating Oil-Cable Sections Using Semi-Stop and Stop Joints *(Courtesy Pirelli Cable Corp.)*

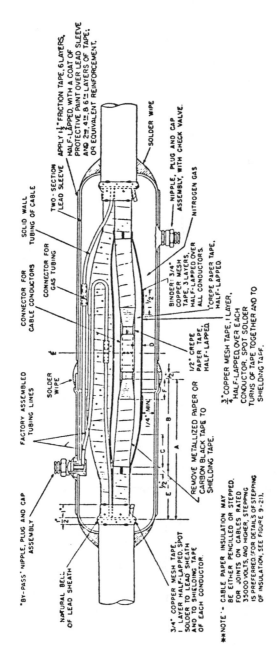

Figure 8-25b. Straight Joint for Gas-Filled Three-Conductor 45 kV Cable. *(From EEI Underground Reference Book)*

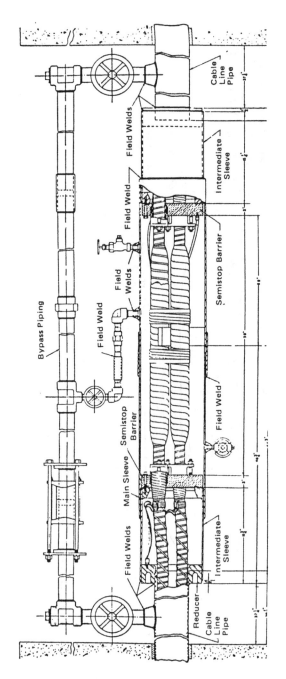

Figure 8-25c. 69 kV Semi-Stop Joint - Welded Casing
(From EEI Underground Reference Book)

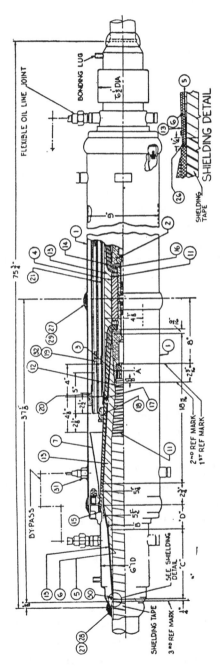

Figure 8-25d. Section of Single Conductor 138 kV Stop Joint

(From EEI Underground Reference Book)

affected section needs to be completely replaced, holding down the time for the repair as well as the cost.

When only a minor repair on an oil-filled cable is required, such as a small oil leak, it may be more expeditious to seal off the small section on which work is to be performed. This may be done by freezing an oil slug from outside the cable or pipe by pouring liquid nitrogen on both sides of the section until the flow of oil is completely stopped. The low temperatures may be maintained by continuing the dripping of liquid nitrogen on ice in a casing placed around the cable or pipe. The cable may then be repaired, pieced out, replaced or rerouted; in the case of pipe cables, the pipe assembly is welded together again. Only the oil in the affected section need be replaced, but it is necessary to ascertain that the oil in the system is free of air, moisture and other contaminants, a long and expensive process. Protective coverings are then reestablished.

Repair or replacement of gas-filled cables is simpler since the gas may be valved off at stop joints or semi-stop joints on both sides of the affected section. The gas in the section between the stop joints or semi-stop joints at which the gas was valved off, is replaced, with the same precautions against moisture and contamination taken for oil, mentioned above.

Extra High Voltage Cable

Transmission voltages above 500 kV are classified as extra high voltage transmission systems. Underground lines of these voltages are usually of the pipe type and the insulation is usually paper impregnated with oil with and in an atmosphere of sulfurhexafluoride gas under pressures of 50 lb/sq in. or higher. Such insulation is capable of withstanding the high voltage electrostatic stresses of such high values under varying temperature and current loading conditions. The high pressure of the gas increases the probability of leakage and intensifies the problems associated with repairs. Underground installations of transmission lines operating at this voltage are generally limited to special circumstances where economics may not be of primary importance.

Superconductors

Conductors whose current-carrying capabilities may be increased substantially by special means are associated with hollow core and pipe type cable installations. Liquefied gas, such as nitrogen, is circulated within the conductors or pipes to maintain an extremely low temperature (cryogenics) which reduces the resistance of the conductors to very low values, decreasing the voltage (IR) drop in it as well as the power loss (I^2R). Aluminum, a common conductor material, if cooled to the temperature of liquid nitrogen (−320°F), has its resistance reduced some 90 percent to approximately one-tenth of its value at normal temperatures, but suffers severe changes in its mechanical properties, losing much of its tensile strength and tending to become brittle. The necessity to maintain the liquidity of the gas for continual removal of heat requires refrigeration and associated equipment of special manufacture and use of certain materials, resulting in extremely expensive installations.

A few comparatively rare metals, Niobium for example, if cooled to temperatures approaching that of liquid helium (−425°F), may lose practically all of its resistance to the flow of direct current. Such conductors are termed superconductors in contrast to the cryogenic conductor mentioned above, the difference being in the materials and temperatures employed.

Despite the difficulties mentioned, economics indicate that cables containing such conductors under the conditions described, may be practical, offering possibilities of substantial economy in the high power capability, but in a (present) range of some 5,000 mVA and above.

Chapter 9

Associated Operations

There are other operations and procedures associated with the transmission and distribution systems that impact on the quality of service rendered the consumer and the economics of electrical design and the efficiency of operations of utility systems.

NON-MANUAL SWITCHING

Earlier the design of transmission and distribution facilities were described that involved the switching of lines to restore service on a faulted circuit by isolating faults and reenergizing the unfaulted parts of the circuit; to transfer loads between circuits or between phases to improve voltage and relieve overloads and potential overloads; to switch on and off capacitors, street lighting and other equipment; to permit deenergization of portions of circuits for construction and maintenance purposes without disturbing the remainder of the circuit; and to perform other operations controlled manually.

With the development of so-called electronic systems for communication and control purposes, coupled with their miniaturization, many of the manual operations may be performed almost instantly by automatic devices actuated by electronic relaying and circuitry. A bit slower, the devices may be operated by radio, controlled from a supervisory source, both with a significant savings in time.

LOAD SHEDDING

The need for load shedding generally arises from unforeseen causes, lack of sufficient power supply from deficiencies in generating, transmission and distribution capabilities. These conditions may stem from unusually higher than foreseen demands that may be due to unusual seasonal changes, special events that may cause loss of diversity, and failure or overload of some elements in the facilities making up the supply system. Occasionally, these conditions may also occur in local areas from load growth unaccompanied with construction of new facilities.

The remedy to these situations is to reduce the demand on the supply bus to match the incoming power available. For relatively small deficiencies, resort may be had to the reduction of voltage on the supply bus. This is because a great part of the demand may consist of lighting (unity power factor) load whose power requirements diminish almost directly with the lowering of the applied voltage; motor power requirements are essentially not affected by the voltage at their terminals. Another means of lowering the demand on the supply bus is the periodic disconnecting of feeders for relatively short periods of time on a predetermined schedule. This is sometimes referred to as a "brownout." Sometimes both these methods may be employed and, if the condition deteriorates, entire substations may be taken out of service temporarily. If conditions continue to worsen where the overloading of vital facilities may cause them to be endangered, an entire area may necessitate a shutdown of the system involved, an operation referred to as "blackout."

Voltage reduction is usually accomplished by controlling the regulators on individual feeders or, as the case may be, on the bus voltage regulator. Voltage may also be reduced by controlling regulators at subtransmission and transmission substations where they may exist; in this case, the regulators on individual feeders or on the distribution substation buses are locked in place to prevent their negating the effect of the voltage reduction on the incoming supply.

Voltage reduction is usually accomplished in steps but may be self-defeating as the light output of lamps may decrease to the point where additional lamps may be turned on; on the other hand, motors may draw more current but continue to operate satisfactorily until the voltage may become too low, torque decreases, they become overheated, and eventually stall.

Care must be exercised in lowering the voltage on feeders supplying low voltage networks. Operation of their regulators must be coordinated so that the load shed by one feeder may not be picked up by the other supply feeders. If not done as simultaneously as possible, the load thus picked up by the feeders whose voltage may not be lowered may cause network protectors to open on the feeder with lowered voltage, and may cause the other feeders to trip from overload, resulting in a cascading effect in which all of the supply feeders may trip, shutting down the network. In some cases, it may be necessary to block the overcurrent relays to prevent the feeders tripping until all of the regulators have been adjusted and locked at the desired voltage level.

Should the network shut down, its re-energization may be accomplished by blocking the overcurrent relays and closing the circuit breakers of the supply feeders as simultaneously as possible; batteries operating the closing mechanisms should be checked to ascertain sufficient output is available for satisfactory operation of the circuit breakers simultaneously. If the network is shut down for any length of time and diversity is lost, it may be necessary to cut the network physically into smaller pieces, blocking open the network protectors on the transformers of pieces of the network so they do not pick up load, picking up a piece of the network at a time. After one piece has been energized, restoring diversity, it may be necessary to open all of the supply feeders, unblock the network protectors on the same feeder in another piece of the network, then reclosing the feeders to pick up the two pieces. The operation may be repeated until all of the pieces are reenergized; the pieces can then be reconnected.

DEMAND CONTROL OR PEAK SUPPRESSION

Economics indicates the desirability of holding down maximum demands or peak loads on the generation, transmission and distribution parts of the electric system. The results include the reduction of investment in plant as well as in operating costs, mainly in fuel, because of reduced I^2R losses. One form, of a temporary nature, is the load shedding procedures described above. Demand control applies to the reduction of maximum demands as a normal ongoing operation.

From an investment point of view, the most effective use of facilities is their operation at maximum loading throughout their lifetime. This would imply their load factors would be 100 percent; this is not always practical, but the higher this value approaches that mark, the better the utilization of the investment and the lower the unit price of the product.

In the supply of a given load, the load factor applies not only to the individual consumer,' but to the entire utility. Typical load factors for consumers may vary from 20 percent for some residential consumers to over 90 percent for some round-the-clock large manufacturing or processing plants; office buildings may have load factors of 20 to 30 percent, and smaller, one shift, industrial and commercial consumers' load factors of from 20 to 70 percent.

By reducing the losses substantially, reflected in lower fuel consumption, demand control becomes a conservation measure that may reduce the call upon the resources of the nation-and the consumer's bill. While the consumer's total overall consumption may remain the same, the reduction of demands will reduce the maximum current flow, even though the reduced current flow may continue for longer periods of time.

As I^2R losses vary as the square of the current, substantial energy savings result. Moreover, experience has shown that measures for reducing demand often reveal the elimination of some unnecessary operations and better methods of operation of equipment, resulting in improved efficiencies and lowered energy costs.

To reduce the demand on its facilities, the utility seeks to reduce individual consumer's demands, but to achieve maximum coincident demand reduction, it is necessary to coordinate the individual consumer demands. This is usually done through rate incentives that specify times of optimum rates for usage at specified off-peak hours. For small residential consumers, resort has been made to radio-controlled devices that meter consumption during peak-load periods for higher rate applications. For large consumers, however, metering to record not only peak demands, but power factor as well, provide incentives for better and overall reduced demands. Indeed, the impulses from the demand meters themselves are used to hold demands on individual pieces of equipment to predetermined values. In reducing the overall demand, analysis of the essentiality of loads is often undertaken and may be categorized as follows:

1. Essential loads that are necessary for safety and operational reasons.

2. Loads that may be curtailed or turned off for short predetermined periods of time, sometimes sequentially, without impairing safety or production.

3. Loads which can be deferred to off-peak times.

In general, load cycling involves the turning on and off of individual loads at predetermined times and often staggered to achieve the smallest maximum demand. Automatic devices have been developed to achieve this result, all based on the consumer's actual consumption and demand, and compared to some predetermined ideal rate of consumption. The different methods employed and their accuracy are not further detailed as they are not within the scope of this work.

LOAD MANAGEMENT CONTROL

Some utilities, in agreement with usually large consumers, have assumed control over devices that automatically switch off and on some of the consumer's loads when undesirable levels are being reached. Noncritical loads, under agreed upon constraints as to the maximum times they may be switched off, are placed under the control of the utility. This may be done by signals transmitted by carrier, radio or telephone. Such "interruptible" load agreements are coupled with favorable rate schedules. Costs for such demand control equipment are sometimes shared with the consumer where the same equipment may be used to control the consumer's maximum demand. Continuous review is usually done with the consumer so that the target values of both parties do not conflict with each other.

UTILITY SUPPLY AND
DEMAND PROBLEMS

Distribution, transmission and generation facilities must be provided to meet daily and seasonal load peaks, maximum demands of relatively short duration. Solutions to these problems are generally based on economics commensurate with service reliability.

DISTRIBUTION

Problems arise as to the loading of conductors, transformers and associated equipment (fuses, switches, regulators, etc.). Although these are nominally rated on current carrying capabilities, actually their limits are based on the allowable temperature at which insulation is safe from failure. The temperature that may pertain does not only result from the heat generated by the losses (copper, iron), but also from the duration of these losses in the unit. The temperature of the ambient also plays a part in the total heat that may affect the condition of the insulation. While the control of demands on the distribution system may reduce the heat generated by the load, the duration of this demand may reduce the thermal margin. The overall effect may be affected appreciably by prolonged ambient temperatures. The units so affected should be closely monitored. The control may, therefore, defer the addition of facilities, not only on the distribution system, but on other facilities back to the generating stations or power sources.

TRANSMISSION

The same observations concerning distribution facilities also apply to transmission circuits. Controlling the demand on these facilities will not only lower the I^2R losses, but also defer, if not obviate, new construction or revamping of transmission lines and equipment. As bulk carriers of electrical energy, transmission lines are, investment-wise, in the same category as generating plants.

GENERATION

Generators on a power system may be classified into three categories: the newest and most efficient base-load units, and usually the most expensive; the most recent of older generators, less efficient but often-operated units; and the least efficient units operated as peak units, generally the oldest requiring much maintenance and including new expensive units specifically designed for short-term peak operations (often gasoline fueled). Controlling system demands results in a lessened need to operate the lesser efficient and the more costly units, and may defer the installation of the most costly newest units.

COGENERATION AND
DISTRIBUTED GENERATION

Changing economic and conservation conditions have made feasible the interconnection of consumer-operated generating facilities to those of the utility through transmission and distribution systems. Indeed, in some areas (e.g., Texas) this has been mandated by law. Large users of steam and hot water who formerly produced their requirements from boilers have found it economically advantageous to generate electricity and use the waste heat for their steam and hot water needs and sell the "by-product" electricity to the local utility, almost always at a profitable rate. Regulatory bodies have favored attractive rates as an inducement to the cogenerators; the rate paid for power purchased by the utility (the avoidable cost to produce power) to be based on the cost of the utility's least efficient generating source, as compared to the cost of power to the utility consumer based on the average cost of power from all the supply sources.

Connecting the cogenerator's generating facilities to the utility's transmission and distribution system requires protective equipment be provided by the cogenerator in accordance with the utility's minimum requirements. The protection includes, but is not limited to, equipment devices that:

1. Synchronize the generator to the utility system automatically, including protection from connecting the cogeneration to the utility system before it can be synchronized.

2. Opens the circuit breaker to disconnect the cogeneration on loss of power in the utility system or when a fault occurs on the utility system tie. Also to provide protection for generator overloading, phase current unbalance, reverse power flow, under and over frequency limits, and under and over voltage limits.

3. Control of the engine governor to regulate speed, loading and phase relationship with the utility system.

The one-line diagram illustrates the electrical connection and the protection involved in the basic, or minimum, requirement for a cogeneration system, Figure 9-1.

The relationship between the utility and cogenerator parallels largely that between utilities participating in a power grid or pool; indeed, cogenerators are essentially other utilities with the same conditions applying, including wheel-barrowing of power and control from a central source (usually the system operator).

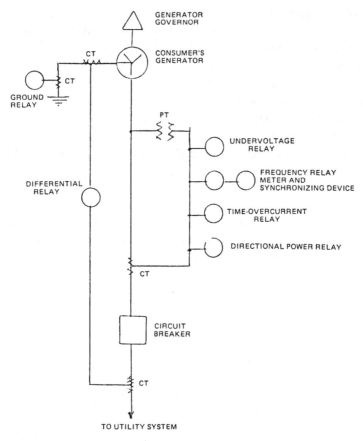

Figure 9-1. One-Line Diagram of Minimum Protective Relaying for Consumer Cogeneration and Distributed Generation Installations

The wide variations in voltage and current distribution in the transmission and distribution circuits to which the cogenerator may be connected may require changes in that circuit configuration to maintain standards of safety and power supply acceptable to the utility. This may require additional facilities to provide adequate sectionalizing and re-energization of circuits, preferably by automatic means. Close coordination of control between the operating groups that control power sources of supply (generation and transmission) and those that control the distribution system, where such separate groups exist, is essential to prevent difficulties from arising.

A smaller version of cogeneration, known as distributed generation is designed to be connected to the distribution system at strategic points. They are used where, for economic or other reasons, they supply loads that would otherwise require additional generating or transmission facilities. They usually consist of small units generally driven by small gas turbines, but may include wind, solar, geothermal, fuel cells and other type units. These may be both utility and consumer owned. These units, and some cogeneration units, are not usually competitive with utility owned larger units that have the advantage of scale.

Some distributed generation (and cogeneration) units may impact negatively on the safety of operations. Although standards for the selection, installation and maintenance of equipment to connect and disconnect these units from the systems to which they supply electric energy, Figure 9-1, are furnished the consumer by the utility, these standards are not always followed, particularly those related to maintenance. This constitutes a hazard to persons who may be working on the systems, believing them to be de-energized, they may be the victims of improper, unannounced connections, energizing the systems to which they are connected. Similarly, should a fault develop on the utility systems to which they are connected, and the equipment fail to disconnect their generation from the system, overloads, fires and explosions may occur. Further, while they are under the supervision of the system operator, they tend to dilute his attention from other events.

METERING

Metering in an electric utility falls into two broad categories: metering for the billing of consumers, that is, for revenue reasons, and metering for operating and monitoring the functions of the several elements of the electric system so as to achieve efficiency and economy in the costs of operation. While these two categories apply to separate and distinct functions, with the advent of digital computers, the data collected for billing purposes may be employed in furthering operating and monitoring purposes.

Billing meters include watt-hour meters, wattmeters or demand meters, meters for measuring reactive power (volt amperes), and power factor for larger commercial and industrial consumers. In the case of larger consumers, the meters may be of both the indicating and recording types, and instrument transformers used to reduce the actual quantities to safe and manageable values. Schematic diagrams of different meter applications are shown in Figure 9-2.

Operational and monitoring meters include, in addition to those mentioned, ammeters, voltmeters, frequency meters, thermometers, barometers, clocks, and indicators and alarm units associated with them. Here, too, metering may be of both recording and indicating types and instrument transformers are almost always employed.

Where cogeneration exists, meters installed may be used for both purposes.

In measuring power in a polyphase circuit, separate wattmeters or watt-hour meters may be connected in each of the phases and the values added algebraically to obtain the total demand or consumption. However, equal accuracy may be obtained by using one less of these meters, the remaining ones being connected properly among the phases of the polyphase circuit. This is known as Blondel's Theorem and may be stated:

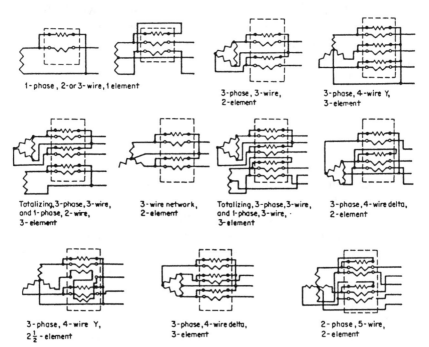

1-phase, 2-or 3-wire, 1 element

3-phase, 3-wire,
2-element

3-phase, 4-wire Y,
3-element

Totalizing, 3-phase, 3-wire,
and 1-phase, 2-wire,
3-element

3-wire network,
2-element

Totalizing, 3-phase, 3-wire,
and 1-phase, 3-wire,
3-element

3-phase, 4-wire delta,
2-element

3-phase, 4-wire Y,
2½-element

3-phase, 4-wire delta,
3-element

2-phase, 5-wire,
2-element

Figure 9-2. Schematic Diagrams of Meter Applications for Different Types of Distribution Systems. *(Courtesy Westinghouse Electric Co.)*

"In any system of N conductors, the true power may be measured by connecting a wattmeter in each conductor but one (N-I wattmeters), the current coil being connected in series in the line and the potential coil connected between that line and the line that has no current coil connected in it; the total power is the algebraic sum of all the readings of the wattmeters so connected."

If the power factor of the circuit should be approximately 50 percent or less, it is probable that one of the wattmeters will register a negative value, in which case it may be necessary to reverse the connections to the terminals of the current or potential coil; should the power factor become greater than the 50 percent, it will be necessary to change back that connection.

REMOTE METER READING AND DEMAND CONTROL

Electronic developments that have made e-mail (and the internet? inexpensive and universal means of communication have also made practical the remote reading of consumer's meters. Periodic inquiry automatically sent to each consumer identifies the meter and records the dial consumption and other data, transmitting it to the computer center where it may be automatically processed, producing the bill sent to the consumer.

In some cases, usually commercial and industrial consumers, where it is desired to hold down their demands by arranging their loads not to coincide, and where practical to be scheduled for off peak hours (usually evening and early morning hours), the same means of communication is used to operate relays and switches to accomplish this purpose, often employing the same reading facilities.

TRANSFORMER LOAD MONITORING

The same computer that translates meter readings into consumers' bills may also be programmed to make selected summaries of such data, simultaneously converting consumption into loads and demands on the several elements of the electric system, but principally on the distribution system. The grid coordinate system of mapping, described in Appendix G, facilitates this collection and conversion of data.

The computer totals the consumption of all of the consumers supplied from one distribution transformer over the billing period of time; factors are applied to these totals to convert them into demands in kW or kVA. Conversion factors are derived and updated periodically by sampling methods. In like manner, the consumption of the transformers on each phase of a distribution circuit may be summarized to determine the maximum demand of the circuit. Applying the same method, the demands on the substation bus and supply transformers may be closely approximated, as well as losses on each feeder. Such data may be used for balancing loads on phases of a circuit, between circuits, and even between substations.

The computer may also be programmed to determine changes in the average demand per consumer, identifying transformers and other elements of the electric system that exceed predetermined values uncovering overloads and potential failures. Hence, damage or destruction of facilities may be averted, improving the reliability of service to consumers and permitting better and more economic planning and design of transmission and distribution systems.

POWER FACTOR CORRECTION

At some large consumers, along with the kilowatt-hour consumption, the reactive kilovolt-ampere hours load is measured (by inserting a reactor, shifting the voltage by 90'), and by properly interrelating these quantities, the average power factor of a consumer over a billing period of time may be determined. Rate schedules may provide rewards or penalties reflecting the consumer's power factor. Correction measures may then be applied, if desirable.

DEMAND CONTROL

This subject has been discussed earlier. The data supplied to the relays or computer supervised machines may be derived from the same meters used for billing purposes.

One, so-called "Ideal Rate," method is illustrated in Figure 9-3. This method depends on the establishment of an ideal rate curve with two offset lines: one to establish a load shed point before the ideal rate is exceeded, and one which allows loads to be restored. Separation between these lines can be adjusted to meet local requirements such as minimum off-on times for equipment. It should be noted that the ideal rate curve does not begin at zero at the start of the demand interval as the "instantaneous rate" curve does. Rather, it starts from an established offset point that takes into account nondiscretionary loads. If this were not done, the lower shed line would keep all equipment off at the beginning of each interval.

The slope of the ideal rate curve is then defined by the offset established for the shed line and the chosen demand set point. The main advantages of the ideal rate method controllers are the ease with which operational modifications can be made to optimize their use (e.g., changing offsets), and their low costs due to restricted computational requirements.

The main disadvantage is that this type of controller must be synchronized with the utility's demand meter to make sure it is controlling over the same interval that the utility is averaging demand. Some utilities may be unwilling to provide the "end of interval" timing pulse that is required to establish this synchronization.

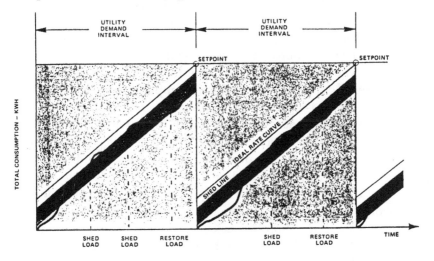

Figure 9-3. "Ideal Rate" Method of Demand Control
(Courtesy McQuay-Perfex Inc.)

RATES

Demand control has also been attempted through a myriad of rates designed to manipulate unit rates depending on a variety of situations: time of year (seasonal); availability of supply, present and proposed (excess or deficiency of capacity, both in generation and transmission as well as capital availability); attraction of industry to area; and other individual or particular situations. For example, in addition to general service rates, there are multiple fuel rates, single fuel rates, primary service rates, block rates, interruptible service rates, and fuel charges, power factor clauses, etc., etc.

In this author's opinion, only consumption of kVA hours should be metered universally. Three simple rates may encompass all the goals presently sought: a promotional rate to encourage greater use; a regular rate to provide for normal growth; and a conservational rate to discourage use and to encourage decrease in peak loads. The first two have been and are in use for the purposes mentioned; it is the third that may be innovative. This would simply call for a reversal of the rates provided for promotional purposes; that is, the greater the consumption, the greater the unit cost.

The rationale for this reverts to the efficiency of the sources of supply. The generators and transmission lines of greatest efficiency serve base load: divide this capacity in kVA by the total number of consumers, or each consumer (of all kinds) is entitled to one share of the most efficient source at its rate of production; the same method then be applied to the next level of source efficiency, dividing only by the number of consumers whose consumption is above that value established for the most efficient source; a third application applied to the least efficient source of supply, using the same method described above. This third rate base could include a possible fourth, one that includes special sources, such as cogeneration, combustion engine-driven peaking units, etc., at the highest rate. The demands for all levels includes the total number of consumers, but the last three levels, divided by only the number of consumers whose consumption is above the levels indicated. This should substantially encourage all to lower their consumption, improve power factor, suppress peaks, etc. Present kW-hour meters may be converted to kVA-hour meters by connecting a suitable reactor in series with the voltage coil; replacement of meters in the high use consumers may be done first. Some interim measures may be devised for correction and uniformity of billing until all meters are converted.

TYPES OF METERS

Simplified wiring diagrams for basic watt-hour meters for use on different types of single and polyphase systems are shown in Figure 9-2.

Electric meters have been developed that measure the quantities discussed earlier with great accuracy and a minimum of maintenance. Adding memory and microcircuitry to the registers enables the simultaneous measurement of several different quantities, such as real and reactive power, peaks, average demands, power factor as well as energy consumption. Also calendar information and complex rate schedules may be introduced. Such systems may also facilitate the remote reading of consumers' meters and instantaneous billing.

TRANSDUCERS

Transducers generally convert nonelectrical quantities into electrical quantities, although they also may reverse this process by converting electrical quantities into nonelectrical ones. Thermocouples, photocells, microphones are examples of devices that convert heat, light and sound into electrical quantities that are utilized in several ways to control or accomplish purposes for which they are intended. Phonographs, telephones, heaters convert electrical quantities into sound, and heat, etc. Other devices convert pressures or differences in pressure into variations of electrical quantities which, in turn, serve to inform or control the condition of equipment with which they may be associated.

NONTECHNICAL FACTORS

In addition to the technical and economic considerations concerning transmission and distribution systems that have been discussed, there are a number of nontechnical (nonelectrical and mechanical) that have a serious impact on the design, construction, maintenance and operation of these facilities.

The foremost consideration is that of safety of employees and the public. Included are the accessibility and nonaccessibility of energized lines and equipment. While electric lines and equipment must be so designed and situated that workers may have safe access to them, they should not interfere with pedestrian and vehicular traffic, nor intrude on such areas where their presence may constitute a particular hazard (e.g., playgrounds, proximity to antennas, flag poles, etc.). Methods, including tools and equipment must provide for safety at all times to the employee and others who may be working on or near energized facilities. Manual or automatic de-energization and grounding of lines and equipment should be considered wherever practical. Clearances, both horizontal and vertical, between energized facilities and adjacent structures should, as a minimum, conform to the NESC.

In the design and construction of lines and equipment, provisions for their replacement, repair, revamping, or adjustment should be included. While de-energization of facilities is recommended, there are occasions where keeping them energized may be required to maintain safety and avoidance of life-threatening situations, so-called live-line and bare-hand methods and procedures may be employed.

Environmental considerations are also of paramount importance. Trees are a particular source of concern. In heavily treed areas, continuity of service may be affected and the proper solution might well be the placing of facilities underground. The same solution may apply where consumers prize the aesthetic value of trees, good community relations may dictate underground construction, particularly if the consumers or public bodies in the area may be willing to share expenses. For appearance sake, overhead lines are routed along rear lot lines, resort is made to armless type construction, substations are landscaped or camouflaged to conform to the environment.

In areas of severe rain, wind, lightning and other natural hazards, facilities may be relocated or placed underground. The same solutions may apply to areas of severe pollution, such as concentrations of salt from spray and chemical contamination. The proximity of airports and military bases may be sources of potential hazard with severe repercussions that relocation or undergrounding may be justified.

In certain cases of community pride, governmental structures, parks, civic centers, historical structures and landmarks, the existence of visible electrical facilities may clash with the environmental spirit intended, and the placing of utility facilities out of sight conforms to the wishes of the community.

Economic considerations must, of course, receive the greatest attention, if the enterprise is to be successful. In addition to the application of Kelvin's Law in the design of transmission and distribution facilities, other factors may be included. Standardization of materials and equipment permits substantial economies to be realized both in their purchase and inventory. It makes for interchangeability and leads to standardization of construction, maintenance and operation practices. In turn, these make employee training more effective as concentration is placed on fewer and repetitive tasks that contribute to the safety and productivity of the work force.

In instances where civic improvement, such as widening of roads, construction or replacement of water and sewer systems, etc., where overhead construction may be normally employed, underground construction may be justified if the costs of relocating facilities, of excavation and restoration, are borne or shared with governmental or other agencies.

To the improvement in tools, equipment, work methods and new techniques in providing facilities to meet the demands of consumers, not only for an ample supply of electricity, but its reliability, must be added provisions for safe and rapid restoration of service in event supply is disturbed or completely cut off. Restoration activities may be expedited, generally by the reduction in elapsed time to remedy the contingency, by improvement in communication, transportation and strategy.

Communication improvement has been rather obvious. Mail, messenger, and telegraph communications have been replaced by telephone and radio, including two-way mobile radio for rapid communication with personnel and crews in the field, and, more recently, the installation of television CRTs (cathode ray tubes) in both field vehicles and operating offices, thus making data in computers rapidly available. Transportation has progressed from horse-drawn ordinary cargo vehicles to specially designed trucks with hydraulically lifted insulated buckets, making the workers' lot easier and safer, and more rapid the completion of restoration work. Vibrating plows and horizontal boring machines replace manual efforts, making practical relatively deep burial of cables, sometimes accomplished by one unit in one operation. Inspection of lines and delivery of personnel and material to transmission job sites for construction and maintenance is speeded by the use of helicopters.

Strategy in the form of prewritten, well thought out procedures, has replaced the unorganized random choice of personnel and equipment. Procedures are periodically revised and updated to reflect trained personnel, new methods and equipment available. Restoration procedures for major contingencies mobilize the resources of not only the utility involved, but other cooperating utilities, contractors, governmental agencies, and others, including public information groups.

Appendix A

Circuit Analysis

INTRODUCTION

In the analysis of circuits to determine current distribution and voltage drops in the individual parts of the circuits, several basic principles are employed. These call for the reduction of the circuitry into successively simpler forms until a single loop circuit results. Computation of current and voltage values can then be made, essentially reversing the order of simplification, until all the individual parts of the circuit have been analyzed. These procedures are especially applicable to network-type circuit.

Kirchhoff s Laws

Kirchhoff's laws encompass two fundamental simple laws which apply to both dc and ac circuits, no matter how complex they may be:

1 . The current flowing away from a point in the circuit (where three or more branches come together) is equal to the amount flowing to that point. Expressed another way, the vector or algebraic sum of all the currents entering (and leaving, a negative entry) a point is zero.

2. The voltage acting between two points in a circuit acts equally on all the paths connected between the two points. Expressed another way, the vector or algebraic sum of all the voltage drops (or rises, negative drops) around a closed loop is zero.

Applying Kirchhoff's laws to the parts of a circuit, a number of equations between the unknowns can be drawn. The number of independent equations which can be written from the first law is 1 less than the number of junction points; from the second law, the number is equal to the number of branches less the number of junction point equations. The equations can be algebraically solved for all the unknowns. In practice, however, these laws are more often used to check results obtained by other means.

CIRCUIT TRANSFORMATIONS

Wye to Delta
Refer to Figure A-1.

$$I_a + I_b + I_c = 0$$
$$I_a = I_B - I_C$$
$$I_b = I_C - I_A$$
$$I_c = I_A - I_B$$
$$\Delta = Z_a Z_b + Z_b Z_c + Z_c Z_a$$

and

$$Z_A = \frac{\Delta}{Z_a} = Z_b + Z_c + \frac{Z_b Z_c}{Z_a} \qquad Z_B = \frac{\Delta}{Z_b} = Z_c + Z_a + \frac{Z_c Z_a}{Z_b}$$

$$Z_c = \frac{\Delta}{Z_c} = Z_a + Z_b + \frac{Z_a Z_b}{Z_c}$$

Delta to Wye
Refer to Figure A-1.

$$I_A = \frac{I_c Z_B - I_b Z_C}{Z_A + Z_B + Z_C} = \frac{(I_c Z_c - I_b Z_b) Z_a}{\Delta'} \qquad Z_A = \frac{Z_B Z_C}{Z_A + Z_B + Z_C}$$

$$I_B = \frac{I_a Z_C - I_c Z_A}{Z_A + Z_B + Z_C} = \frac{(I_a Z_a - I_c Z_c) Z_b}{\Delta'} \qquad Z_b = \frac{Z_C Z_A}{Z_A + Z_B + Z_C}$$

$$I_A = \frac{I_b Z_A - I_a Z_B}{Z_A + Z_B + Z_C} = \frac{(I_b Z_b - I_a Z_a) Z_c}{\Delta i} \qquad Z_C = \frac{Z_A Z_B}{Z_A + Z_B + Z_C}$$

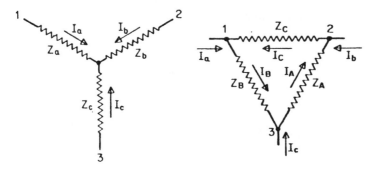

Figure A-1. Delta-Wye Transformations

where $\Delta' = Z_A Z_B + Z_B Z_C + Z_C Z_A$.

Changing Bases

To convert Z in ohms at a voltage E to Z', the equivalent value on a voltage base E',

$$Z' = A\left(\frac{E'}{E}\right)^2$$

To convert Z in percent on a kVA base U to Z, the equivalent value on a kVA base U',

$$Z' = Z\frac{U'}{U}$$

To convert Z_p in percent on a kVA base U to Z, in ohms on a voltage base E,

$$Z_z = \frac{3Z_p E^2}{U \times 10^5}$$

and, conversely

$$Z_p = \frac{U \times 10^5}{3E^2}$$

where E = line-to-neutral voltage and U total three-phase kVA.

Paralleling Two Impedances
    Refer to Figure A-2.

$$Z = \frac{Z_a Z_b}{Z_a + Z_b}$$

where Z in the equivalent impedance: and

$$I_a = I\frac{Z_b}{Z_a + Z_b} \qquad I_b = I\frac{Z_a}{Z_a + Z_b}$$

SUPERPOSITION THEOREM

    In a network containing several voltage sources, the current in the several branches may be found by replacing all but one of the voltage sources by their particular resistances (dc) or impedances (ac) and determining the current contributed by the one source in each of the branches. The process is repeated with each of the other voltage sources, and separate current distribution in the several branches from each of the voltage sources is again determined, The vector or algebraic sum of all of the currents in each branch (as determined above) gives the value of the current in that branch with all of the voltage sources in place.

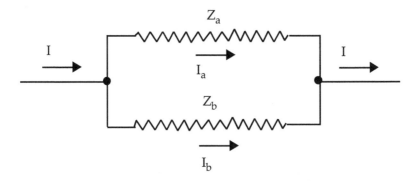

Figure A-2. Paralleling Two Impedances

Thevenin's Theorem—The current in any terminating impedance Z_T connected to any network is the same as if Z_R were connected to a generator whose voltage is the open circuit voltage of the network, and whose internal impedance Z_R is the impedance looking back from the terminals of Z_T, with all generators replaced by impedances equal to the internal impedance of these generators.

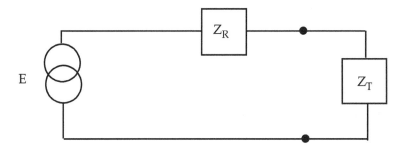

Figure A-3. Thevenin's Theorem

Norton's Theorem—The current in any terminating impedance Z_T connected to any network is the same as if Z_T were connected to the parallel combination of a current generator whose current is equal to that delivered by the actual network to a short circuit connection across the terminals Z_T and an impedance of Z_R which is the impedance looking back from the terminals of Z_T with all the generators replaced by impedances equal to the internal impedance of these generators.

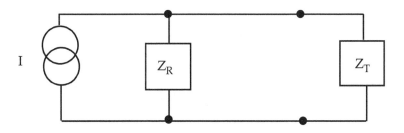

Figure A-4. Norton's Theorem

Appendix B*

Symmetrical Components

C. F. WAGNER and R. D. EVANS
Transmission Engineers,
Westinghouse Electric & Mfg. Co.

THE SOLUTION of asymmetrical polyphase networks by the method of symmetrical components was discovered by Mr. C. L. Fortescue, the first publication appearing in 1918. Since then the method has been applied in the solution of a wide variety of engineering problems. It has been used most extensively in electrical engineering for the solution of problems involving unsymmetrical load or circuit conditions.

As an example of the applicability of this method consider the case of faults to ground on a power system. The ordinary methods of calculation, being based on symmetrical voltages and currents, are therefore limited to the case of the three-phase short-circuits, whereas it is probable that from 95 to 99 percent of all faults originate single-phase, either from line to ground or from line to line. The method of symmetrical components, however, is particularly well adapted to the calculation

---

*(Courtesy Westinghouse Electric Co.)

of single-phase faults. In fact it is the only practical method of determining the voltages and currents in the different phases in all parts of a power system, under conditions of single-phase load or short-circuit.

A knowledge of the method of symmetrical components is essential for an adequate understanding of application problems involving the magnitude and phase relation of voltages and circuits under unbalanced conditions, and is therefore necessary for the design of power systems from the standpoint of circuit-breaker application, relay protection, and stresses in electrical machinery. The method is particularly suitable for the analysis of the performance of rotating machinery for single-phase or unbalanced polyphase operation, and has been used quite extensively in connection with the design of machinery on single-phase railway systems, particularly from the standpoint of phase-converting apparatus. A number of new schemes have appeared commercially which make use of conceptions arising from the method of symmetrical components, such as the negative-sequence protective relay and the positive-sequence voltage regulator. In recent years stability of power systems has been recognized as an important problem, and it has been shown that the limiting condition arises at times of faults, hence the method of symmetrical components has been of very considerable assistance in the analysis of stability problems.

The principles underlying the subject of symmetrical components were first presented in a paper by Mr. C. L. Fortescue before the American Institute of Electrical Engineers. Mr. Fortescue's treatment is quite general and thorough, and covers a large field in illustrating the application of the method to different types of problems. It necessitates, however, a familiarity with certain branches of higher mathematics which, through lack of application, the average engineer has forgotten. For this reason, and in response to repeated demands for a more popular presentation of the subject, the authors have undertaken this task.

The plan is to review briefly the laws governing complex numbers and then to develop certain theorems regarding the resolution of simple complex numbers into symmetrical groups. In particular these developments will apply largely to electrical systems involving symmetrical or balanced impedances. Illustrations of these theorems will be made by applying them to practical cases, such as the analysis of short-circuits in transmission networks. This will be followed by the application to a few miscellaneous problems, some of which will involve unsymmetrical impedances.

The treatment will be limited to three-phase systems only, except that the last article will develop some of the more important relations as applied to two- or four-phase systems. The authors have drawn liberally from the unpublished as well as the published work of Messrs. Fortescue, Peters, and Slepian, and gratefully acknowledge the inspiration derived from personal contact with these men.

LAWS OF COMPLEX NUMBERS

Any complex number, such as a + jb, may be represented by a single point plotted on Cartesian coordinates, in which a is the abscissa and b the ordinate. This is illustrated in Fig. B-1. It is often convenient, however, to represent complex numbers by another method. Referring to Fig. B-1, let r represent the length of the line connecting the point with the origin and θ the angle measured from the X axis to the line r. It is apparent that,

$$a = r \cos \theta$$
$$b = r \sin \theta$$

or $a + jb = r(\cos\theta + j \sin\theta)$ .... (1)

This is called the polar form, in which θ is the argument or amplitude and r the modulus or absolute value. Increasing θ in a positive sense rotates OP counterclockwise. Similarly,

$$a - jb = r(\cos\theta - j \sin\theta) \dots \dots \dots \dots \dots (2)$$

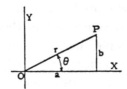

Fig. B-1. A Complex Number

Increasing θ in a positive sense in this case produces clockwise rotation.

If ϵ^θ be expanded into an infinite series,

$$\epsilon^\theta = 1 + \theta + \frac{\theta^2}{2!} + \frac{\theta^3}{3!} + \frac{\theta^4}{4!} + \frac{\theta^5}{5!} + \frac{\theta^6}{6!} + \ldots\ldots$$

Similarly,

$$\epsilon^{j\theta} = 1 + j\theta - \frac{\theta^2}{2!} - j\frac{\theta^3}{3!} + \frac{\theta^4}{4!} + j\frac{\theta^5}{5!} - \frac{\theta^6}{6!} \ldots\ldots$$

Separating the real and imaginary components,

$$\epsilon^{j\theta} = \left(1 - \frac{\theta^2}{2!} + \frac{\theta^4}{4!} - \frac{\theta^6}{6!} + \ldots\ldots\right)$$

$$+ j\left(\theta - \frac{\theta^3}{3!} + \frac{\theta^5}{5!} - \frac{\theta^7}{7!} + \ldots\ldots\right)$$

These expansions* are equal to $\cos\theta$ and $\sin\theta$ respectively, so that,

$$\epsilon^{j\theta} = \cos\theta + j\sin\theta \quad\ldots\ldots\ldots\ldots\ldots\ldots\ldots\ldots\ldots (3)$$

---

*These relations may be verified by reference to any standard college algebra.

Similarly,

$$\epsilon^{-j\theta} = \cos\theta - j\sin\theta \dots\dots\dots\dots\dots\dots (4)$$

This suggests another form to represent a complex number. Substituting (3) in (1) and (4) in (2),

$$a + jb = r\epsilon^{j\theta} \dots\dots\dots\dots\dots\dots\dots\dots (5)$$
$$a - jb = r\epsilon^{-j\theta} \dots\dots\dots\dots\dots\dots\dots\dots (6)$$

These forms will be found valuable for operations involving multiplication and division.

Returning to equations (1) and (2) and (5) and (6) it will be noted that the real components are equal, but that the imaginary components have opposite signs. These numbers are called conjugates.

In the notation used in this series of articles an ordinary complex number will be represented by placing the symbol ˅ above the quantity, for example Ǐ. The conjugate will be designated by the inverse, Î. This notation may be easily remembered by associating the symbol with the first letter of the word vector.

$$\text{If } \check{I} = a + jb \text{ the conjugate will be,}$$
$$\hat{I} = a - jb$$

The real and imaginary components may be expressed in terms of the complex number and its conjugate as follows:

$$a = \frac{1}{2}(\check{I} + \hat{I}) \dots\dots\dots\dots\dots\dots\dots\dots (7)$$

$$jb = \frac{1}{2}(\check{I} - \hat{I}) \dots\dots\dots\dots\dots\dots\dots\dots (8)$$

The law for the addition of exponents holds equally well for both real and imaginary quantities. Just as $10^3 \times 10^4 = 10^7$ so,

$$\epsilon^{j\theta_1} \times \epsilon^{j\theta_2} = \epsilon^{j(\theta_1 + \theta_2)}$$

from which it follows that the product of two complex numbers,

$$\check{E} = E \, e^{j\theta_1} \text{ and } \check{I} = I \, e^{j\theta_2}$$

is equal to,

$$\check{E}\check{I} = EI \, e^{j(\theta_1 + \theta_2)} \quad \dots\dots\dots\dots\dots\dots\dots\dots \quad (9)$$

It will be noted that the absolute value of the product is the product of the absolute values of the components and that the argument is the sum of the argument.

Multiplying by $e^{j\theta}$ merely rotates the modulus through an angle θ in a counterclockwise direction. This point is important as it forms the basis for the proof of quite a number of circle diagrams.

If a complex number be multiplied by the conjugate of another, the argument of the product is the difference in arguments of the components.

$$\check{E}\hat{I} = E \, e^{j\theta_1} \, I \, e^{j-\theta_2}$$

$$= EI \, e^{j(\theta_1 - \theta_2)} \quad \dots\dots\dots\dots\dots\dots\dots\dots \quad (10)$$

Resolved into real and imaginary components,

$$\check{E}\hat{I} = EI[\cos(\theta_1 - \theta_2) + j \sin(\theta_1 - \theta_2)] \quad \dots\dots\dots \quad (11)$$

Squaring a complex number merely increases the modulus as the square and rotates it through double the angle.

$$(I \, e^{j\theta})^2 = I^2 \, e^{j2\theta} \quad \dots\dots\dots\dots\dots\dots\dots\dots \quad (12)$$

Division, the inverse of multiplication, can be accomplished in a similar manner.

$$\frac{\check{E}}{\check{I}} = \frac{E e^{j\theta_1}}{I \, e^{j\theta_2}} = \frac{E}{I} \, e^{j(\theta_1 - \theta_2)} \quad \dots\dots\dots\dots\dots\dots \quad (13)$$

The absolute value of the resultant is equal to the quotient of the absolute values, and the argument is the difference of the arguments of the components.

As squaring a number merely increases the modulus to the square of the number and doubles the argument, extracting the square root, the inverse, is accomplished by taking the square root of the modulus and halving the argument. Generalizing, the modulus of the nth root of a number is the nth root of the modulus of the number and the argument is $\dfrac{1}{n}$ of the argument of the number.

$$\sqrt[n]{I\,e^{j\theta}} = \sqrt[n]{I}\,e^{j\frac{\theta}{n}} \dots\dots\dots\dots\dots\dots\dots\dots\dots\dots\dots (14)$$

If a complex number be multiplied by its conjugate, the product is a real number of magnitude equal to the square of the absolute value of the number, for example,

$$\check{I}\,\hat{I} = I\,e^{j\theta}\,I\,\epsilon^{-j\theta}$$

$$= I^2\,e^{j(\theta-\theta)}$$

$$= I^2 \dots\dots\dots\dots\dots\dots\dots\dots\dots\dots\dots\dots\dots (15)$$

VECTOR REPRESENTATION OF ALTERNATING QUANTITIES

The instantaneous value of a simple harmonic function, such as an alternating electromotive force, may be represented by the following equation:

$$e = \sqrt{2}\,E\,\cos\,(\omega\,t + \alpha)$$

where
$$\omega = 2\,\pi\,f$$
$$f = \text{frequency}$$
$$E = \text{r.m.s. magnitude}$$
$$\sqrt{2}\,E = \text{crest value}$$
$$\alpha = \text{angle which determines the value of } e \text{ for } t = 0.$$

By the addition of equations (3) and (4) it may be seen that $\cos\theta = \dfrac{1}{2}(e^{j\theta} + \epsilon^{-j\theta})$ and substituting $\omega t + \alpha$ for θ, gives,

$$\sqrt{2}\, E \cos(\omega t + \alpha) = \sqrt{2}\, E \left[\frac{\epsilon^{j(\omega t + \alpha)} + \epsilon^{-j(\omega t + \alpha)}}{2} \right]$$

$$= \sqrt{2}\, E \left[\frac{\epsilon^{j\alpha}\epsilon^{j\omega t} + \epsilon^{-j\alpha}\epsilon^{-j\omega t}}{2} \right]$$

So that, if we let,

$$\sqrt{2}\, E\, \epsilon^{j\alpha} = \sqrt{2}\, \check{E} \quad \text{and} \quad \sqrt{2}\, E\, \epsilon^{-j\alpha} = \sqrt{2}\, \hat{E}$$

$$e = \frac{\sqrt{2}}{2} (\check{E}\, \epsilon^{j\omega t} + \hat{E}\, \epsilon^{-j\omega t}) \ldots \ldots \ldots \ldots (16)$$

It will be noted from this equation that the instantaneous value can be represented as the sum of two oppositely rotating vectors which are conjugate at any instant, the imaginary components cancelling out. This relation is shown in Fig. B-2, in which the full lines indicate the positions of $\check{E}$ and $\hat{E}$ for $t = 0$ and the dotted lines at an instant later. It will be noted that the two vectors are always symmetrically disposed relative to the X-axis so that their sum is always real and does not contain an imaginary component. Furthermore, the projection of either vector on the X-axis is always equal to one-half the sum of the vectors, and the instantaneous value can be represented by the real component of either vector. The same relations apply equally well to the instantaneous values of current.

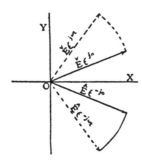

Fig. B-2. A Simple Harmonic Function

VECTOR REPRESENTATION
OF POWER

By instantaneous power is meant the product of the instantaneous values of current and voltage. When the latter quantities vary in a sinusoidal manner, it may be shown that the expression for instantaneous power consists of a constant term with a sinusoidal term of double frequency superimposed. However, in most calculations one is interested only in the average value. It will be shown that this value can be determined in terms of the vector values of current and voltage in the form $\check{E}\hat{I}$ or $\hat{E}\check{I}$. These expressions are of particular value in analytical work, not only because they furnish a convenient expression for average power but because the imaginary component represents reactive power. To show these important relations let the instantaneous value of voltage be

$$e = \frac{\sqrt{2}}{2}(\check{E}\,\epsilon^{j\omega t} + \hat{E}\,\epsilon^{-j\omega t}) \dots\dots\dots\dots\dots\dots \ (17)$$

and the instantaneous value of current be

$$i = \frac{\sqrt{2}}{2}(\check{I}\,\epsilon^{j\omega t} + \hat{I}\,\epsilon^{-j\omega t}) \dots\dots\dots\dots\dots\dots \ (18)$$

The instantaneous value of power is then

$$ei = \frac{(\check{E}\hat{I} + \check{E}\check{I}\,\epsilon^{j2\omega t})}{2} + \frac{(\hat{E}\check{I} + \hat{E}\hat{I}\,\epsilon^{-j2\omega t})}{2} \dots\dots\dots \ (19)$$

The two parts on the right-hand side are conjugate to each other. $\check{E}\hat{I}$ and $\hat{E}\check{I}$ are conjugate vectors of equal magnitude independent of time, lying in opposite quadrants. The double frequency rotational vectors $\check{E}\check{I}\,\epsilon^{j2\omega t}$ and $\hat{E}\hat{I}\,\epsilon^{-j2\omega t}$ are also conjugate to each other and have the same absolute values as $\check{E}\hat{I}$ and $\hat{E}\check{I}$. These relations are illustrated in Fig. B-3.

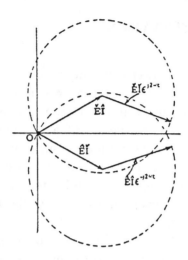

Fig. B-3. Double-Frequency Rotating Vectors

It can be seen that the average values of the double-frequency components are zero so that the mean value of power P is equal to $\frac{1}{2}(\check{E}\,\hat{I} + \hat{E}\,\check{I})$.

Now if $\hat{E} = E\,e^{j\alpha}$ and $\hat{I} = I\,e^{j\beta}$ then

$$P = \frac{1}{2}\left[E\,I\,e^{j(\alpha-\beta)} + E\,I\,e^{-j(\alpha-\beta)}\right]$$

$$= \frac{E\,I}{2}\left[\{\cos(\alpha-\beta) + j\sin(\alpha-\beta)\} + \{\cos(\alpha-\beta) - j\sin(\alpha-\beta)\}\right]$$

$$= E\,I\cos(\alpha-\beta) \quad\dots\dots\dots\dots\dots\dots\dots\dots \text{(20)}$$

This is the familiar form of expression that is usually used. It will be observed that the real parts of both $\check{E}\,\hat{I}$ or $\hat{E}\,\check{I}$ are equal to this value and from this point of view, having once defined P as the real part of $\check{E}\,\hat{I}$ or $\hat{E}\,\check{I}$, either expression can be used.

A rigorous proof for a similar relation for reactive power is quite involved, but starting with the definition of reactive power as equal to E I sin $(\alpha - \beta)$ it is possible to show that the imaginary parts of the above expressions $\check{E}\hat{I}$ and $\hat{E}\check{I}$ equal the reactive power. To do this we must first define positive reactive power. Mr. Fortescue, in first presenting these forms, preferred to call reactive power positively when the current lags the voltage. This assumption immediately fixes the sense in which the angle $(\alpha - \beta)$ is measured, and determines that $(\alpha - \beta)$ shall be positive when the current lags the voltage, and negative when the current leads the voltage.

The imaginary component of $\check{E}\hat{I}$ is E I sin $(\alpha - \beta)$ but the imaginary component of $\hat{E}\check{I}$ is $-$ E I sin $(\alpha - \beta)$. The former is, therefore, of proper sign to conform with the original assumption of reactive power. Both real and reactive power may therefore be determined by the expression

$$P + jQ = \check{E}\hat{I} \dots\dots\dots\dots\dots\dots\dots\dots\dots\dots\dots (21)$$

However, had we started with the definition that positive reactive power conformed to leading current then we should have obrained the result

$$P + jQ = \hat{E}\check{I} \dots\dots\dots\dots\dots\dots\dots\dots\dots\dots\dots (22)$$

The choice of either expression is, therefore, dependent entirely upon the original assumption. It is important that, after having made a choice, one should be consistent throughout the particular problem. In these articles the expression $\check{E}\hat{I}$ will be used.

RESOLUTION OF THREE VECTORS INTO SYMMETRICAL COMPONENTS

Any point on a line is completely defined by the distance from a given reference point, either in the positive direction or the negative. Similarly a point on a plane is completely defined by the distance from two reference axes and a point in space by the distance from three reference axes. A point on a line, being restrained to lie in the line, is said to possess but one degree of freedom; the point in the plane two degrees of freedom; and the point in space three degrees of freedom. So also with a co-planar vector, since it is determined by the position of its terminal it is said to possess two degrees of freedom.

Similar considerations apply to systems of vectors. For example, a system of three co-planar vectors is completely defined by six parameters; the system possesses six degrees of freedom. When, however, one applies the restriction that the system be symmetrical, the added restraint reduces the system to one of two degrees of freedom. Going a step further, it is quite conceivable that a system of three co-planar vectors with six degrees of freedom can be defined in terms of three symmetrical systems of vectors each having two degrees of freedom.

Consider first the symmetrical system of vectors $\check{E}_{a1}$, $\check{E}_{b1}$ and $\check{E}_{c1}$ in Fig. B-4(a). Being balanced, the vectors have equal amplitudes, are displaced 120 degrees relative to each other and,

$$\left.\begin{array}{l} \check{E}_{a1} = \check{E}_{a1} \\ \check{E}_{b1} = \epsilon^{j240}\,\check{E}_{a1} = a^2\,\check{E}_{a1} \\ \check{E}_{c1} = \epsilon^{j120}\,\check{E}_{a1} = a\,\check{E}_{a1} \end{array}\right\} \quad \dots\dots\dots\dots\dots\dots (23)$$

in which a is the unit vector, $\epsilon^{j120} = -0.5 + j\,0.866$

and a^2 is the unit vector, $\epsilon^{j240} = -0.5 - j\,0.866$ $\left.\right\}$ .... (24)

Because of the frequent use of this special vector the symbol denoting a vector will be omitted and this particular letter reserved for this use only. Some of the properties of this vector are given in Table B-1.

This system of vectors is called the positive sequence system, because $\check{E}_{a1}$ leads $\check{E}_{b1}$ by 120 degrees and $\check{E}_{b1}$ leads $\check{E}_{c1}$ by 120 degrees. Their instantaneous values pass through zero in the order $\check{E}_{a1}$, $\check{E}_{b1}$, $\check{E}_{c1}$. The system must always be treated as a unit. In a three-phase balanced electrical system E_{a1} would represent the voltage of phase A; $a^2\,\check{E}_{a1}$ the voltage of phase B; and $a\,\check{E}_{a1}$ the voltage of phase C.

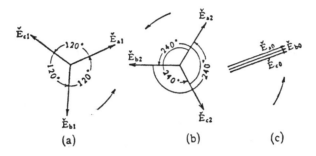

Fig. B-4. Vector Systems

Table B-1. Properties of the Vector a

$$1 = \epsilon^{j0} = 1.0 + j\,0.0$$
$$a = \epsilon^{j120} = -0.5 + j\,0.866$$
$$a^2 = \epsilon^{j240} = -0.5 - j\,0.866$$
$$a^3 = \epsilon^{j360} = \epsilon^{j0} = 1.0 + j\,0.0$$
$$a^4 = \epsilon^{j480} = \epsilon^{j120}$$
$$a^5 = \epsilon^{j600} = \epsilon^{j240}$$

$$1 + a^2 + a = 0$$
$$a - a^2 = \sqrt{3}\,\epsilon^{j90} = j\sqrt{3}$$
$$a^2 - a = \sqrt{3}\,\epsilon^{-j90} = -j\sqrt{3}$$
$$1 - a = \sqrt{3}\,\epsilon^{-j30} = 1.5 - j\,0.866$$
$$1 - a^2 = \sqrt{3}\,\epsilon^{j30} = 1.5 + j\,0.866$$

Likewise, consider the balanced system of vectors $\check{E}_{a2}$, $\check{E}_{b2}$, and $\check{E}_{c2}$ in Fig. B-4(b).

$$\left. \begin{aligned} \check{E}_{a2} &= \check{E}_{a2} \\ \check{E}_{b2} &= \epsilon^{j120}\,\check{E}_{a2} = a\,\check{E}_{a2} \\ \check{E}_{c2} &= \epsilon^{j240}\,\check{E}_{a2} = a^2\,\check{E}_{a2} \end{aligned} \right\} \quad \dots\dots\dots\dots\dots (25)$$

This system of vectors is called the negative sequence system.

A physical picture of the significance of these systems may be obtained by considering the fields which result when voltages of the two systems are applied to the windings of a three-phase machine with distributed windings, such as an induction motor. If the a, b, and c phases of the positive sequence voltages be applied to the terminals a, b, c, respectively, a magnetic field will be produced which will revolve in a certain direction. If now the a, b, and c phases of the negative sequence voltages be connected to the a, b, and c terminals a magnetic field revolving in the opposite direction will be produced. This same effect may be produced by reversing one pair of leads but it will be noted that this in reality is the same as changing the phase sequence from abc to acb where the order indicates the order in which the different phases pass through zero.

Phase sequence should not be confused with the rotation of the vectors. In this series of articles the standard notion of counterclockwise rotation will be accepted for all vectors; both positive and negative sequence vectors will rotate in the same direction but the resultant fields in machines will have opposite rotations.

And finally, consider the system $\overset{\smile}{E}_{a0}$, $\overset{\smile}{E}_{b0}$, and $\overset{\smile}{E}_{c0}$ in Fig. B-4(c) which represents three equal vectors.

$$\left.\begin{array}{l} \overset{\smile}{E}_{a0} = \overset{\smile}{E}_{a0} \\[4pt] \overset{\smile}{E}_{b0} = \overset{\smile}{E}_{a0} \\[4pt] \overset{\smile}{E}_{c0} = \overset{\smile}{E}_{a0} \end{array}\right\} \quad \dots\dots\dots\dots\dots\dots\dots\dots\dots (26)$$

This system is called the zero sequence system.

In all three systems the suffixes denote the components of the respective systems in the different phases. The total voltage of any phase is then equal to the sum of the particular components of the different sequences in that phase. Therefore, any three arbitrary vectors $\overset{\smile}{E}_a$, $\overset{\smile}{E}_b$, and $\overset{\smile}{E}_c$ may be equated,

$$\overset{\smile}{E}_a = \overset{\smile}{E}_{a0} + \overset{\smile}{E}_{a1} + \overset{\smile}{E}_{a2}$$

$$\check{E}_b = \check{E}_{b0} + \check{E}_{b1} + \check{E}_{b2}$$
$$\check{E}_c = \check{E}_{c0} + \check{E}_{c1} + \check{E}_{c2}$$

or substituting their equivalent values,

$$\check{E}_a = \check{E}_{a0} + \check{E}_{a1} + \check{E}_{a2} \quad\dots\dots\dots\dots\dots\dots\dots\dots \quad (27)$$

$$\check{E}_b = \check{E}_{a0} + a^2\,\check{E}_{a1} + a\,\check{E}_{a2} \quad\dots\dots\dots\dots\dots\dots \quad (28)$$

$$\check{E}_c = \check{E}_{a0} + a\,\check{E}_{a1} + a^2\,\check{E}_{a2} \quad\dots\dots\dots\dots\dots\dots \quad (29)$$

The arbitrary system is thus defined in terms of three balanced systems. The individual components may be obtained as follows:

Adding the three equations, and recalling that $1 + a^2 + a = 0$

$$\check{E}_{a0} = \frac{\check{E}_a + \check{E}_b + \check{E}_c}{3} \quad\dots\dots\dots\dots\dots\dots\dots\dots \quad (30)$$

Multiplying equation (28) by a and equation (29) by a^2 and adding,

$$\check{E}_{a1} = \frac{\check{E}_a + a\,\check{E}_b + a^2\,\check{E}_c}{3} \quad\dots\dots\dots\dots\dots\dots \quad (31)$$

Multiplying equation (28) by a^2 and equation (29) by a and adding,

$$\check{E}_{a2} = \frac{\check{E}_a + a^2\,\check{E}_b + a\,\check{E}_c}{3} \quad\dots\dots\dots\dots\dots\dots \quad (32)$$

Suppose the given voltages are balanced, for example,

$$\check{E}_a = \check{E}_a$$
$$\check{E}_b = a^2\,\check{E}_a$$
$$\check{E}_c = a\,\check{E}_a$$

Then,
$$\check{E}_{a0} = \frac{(1 + a^2 + a)}{3}\,\check{E}_a = 0$$

$$\breve{E}_{a1} = \frac{(1 + a^3 + a^3)}{3} \breve{E}_a = \breve{E}_a$$

$$\breve{E}_{a2} = \frac{1 + a^4 + a^2}{3} \breve{E}_a = 0$$

The zero and negative sequence components of voltage in a balanced three-phase system disappear, leaving only the positive sequence.

The graphical construction for the determination of the sequence components will aid further in elucidating the operations. In Fig. B-5 are shown three vectors, $\breve{E}_a$, $\breve{E}_b$, $\breve{E}_c$. The zero sequence component $\breve{E}_{a0}$ is obtained by direct addition. The positive sequence component $\breve{E}_{a1}$ is obtained by rotating $\breve{E}_b$ counterclockwise 120 degrees, $\breve{E}_c$ counterclockwise 240 degrees, and taking one-third of the sum. Similarly $\breve{E}_{a2}$ is determined by rotating $\breve{E}_b$ clockwise 120 degrees, $\breve{E}_c$ counterclockwise 120 degrees, and taking one-third of the sum.

The method may be further illustrated by means of the analytical calculation of the components. Consider the three vectors,

$$\breve{E}_a = 60 + j\,0$$
$$\breve{E}_b = 45 - j\,75$$
$$\breve{E}_c = -21 + j\,120$$

From equations (30), (31) and (32),

$$\breve{E}_{a0} = \frac{1}{3}\left[\breve{E}_a + \breve{E}_b + \breve{E}_c\right]$$

$$= \frac{1}{3}\left[(60 + j\,0) + (45 - j\,75) + (-21 + j\,120)\right]$$

$$= 28 + j\,15$$

$$\breve{E}_{a1} = \frac{1}{3}\left[\breve{E}_a + a\,\breve{E}_b + a^2\,\breve{E}_c\right]$$

$$= \frac{1}{3}\left[(60 + j\,0) + (-0.5 + j\,0.866)\right.$$

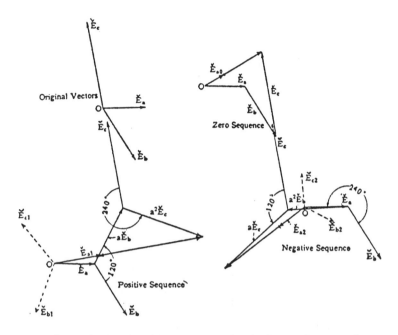

**Fig. B-5. Graphical Construction for the Determination of
Sequence Components**

$$(+\,45 - j\,75) + (-0.5 - j\,0.866)\,(-21 + j\,120)$$

$$= 72.2 + j\,11.5$$

$$\breve{E}_{a2} = \frac{1}{3}\left[\breve{E}_a + a^2\,\breve{E}_b + a\,\breve{E}_c\right]$$

$$= \frac{1}{3}\left[(60 + j\,0) + (-0.5 - j\,0.866)\right.$$

$$\left. (45 - j\,75) + (-0.5 + j\,0.866)\,(-21 + j\,120)\right]$$

$$= (-40.2 - j\,26.5)$$

These three values give the symmetrical components of the
three original vectors. Using these same values of components
the determination of the original vectors in terms of their components may be illustrated.

From equations (27), (28) and (29),

$$\check{E}_a = \check{E}_{a0} + \check{E}_{a1} + \check{E}_{a2}$$

$$= (28 + j\,15) + (72.2 + j\,11.5) + (-40.2 - j\,26.5)$$

$$= 60 + j\,0$$

$$\check{E}_b = \check{E}_{a0} + a^2\check{E}_{a1} + a\,\check{E}_{a2}$$

$$= (28 + j\,15) + (-0.5 - j\,0.866)\,(72.2 + j\,11.5)$$

$$+ (-0.5 + j\,0.866)\,(-40.2 - j\,26.5)$$

$$= 45 - j\,75$$

$$\check{E}_c = \check{E}_{a0} + a\,\check{E}_{a1} + a^2\,\check{E}_{a2}$$

$$= (28 + j\,15) + (-0.5 + j\,0.866)\,(72.2 + 11.5)$$

$$+ (-0.5 - j\,0.866)\,(-40.2 - j\,26.5)$$

$$= -21 + j\,120$$

SEQUENCE OPERATOR

Many mathematical operations may be designated by simplified notations, e.g. $\sqrt{}$ indicates that the square root of the number under the radical must be taken, $\dfrac{d}{dt}$ indicates that the function following should be differentiated with respect to time. These symbols are called operators and constitute shorthand methods of indicating certain mathematical operations. The j in the complex algebra of alternating quantities may also be considered as an operator. It rotates the vector which is operated upon through 90 degrees. The operations in the three sequences may be indicated in a similar manner by means of the sequence operator. This operator is particularly useful for problems involving unsymmetrical impedances. Let S^1 $(\check{E}_{a1})$ represent the three vectors $\check{E}_{a1}$, $a^2 \check{E}_{a1}$, $a \check{E}_{a1}$. S^1 may be considered as an operator affecting the three vectors $\check{E}_{a1}$, $\check{E}_{a1}$, $\check{E}_{a1}$, simultaneously by multiplying the first by 1, the second by a^2 and the third by a. It is an operator that can operate only on a polyphase system.

S^2 $(\check{E}_{a2})$ may represent the system of vectors, $\check{E}_{a2}$, $a \check{E}_{a2}$, $a^2 \check{E}_{a2}$ and the operator S^2 shall be taken to mean that the polyphase vectors $\check{E}_{a2}$, $\check{E}_{a2}$, and $\check{E}_{a2}$ are multiplied respectively by 1, a and a^2.

S^0 $(\check{E}_{a0})$ shall represent the system of vectors $\check{E}_{a0}$, $\check{E}_{a0}$, $\check{E}_{a0}$ and the operator S^0 the simultaneous multiplication of the three vectors $\check{E}_a$, $\check{E}_a$, and $\check{E}_a$ by unity.

So we may write,

$$S^0 \, (\check{E}_{a0}) = (1, 1, 1)\check{E}_{a0} \quad \dots\dots\dots\dots\dots\dots\dots \quad (33)$$

$$S^1 \, (E_{a1}) = (1, a^2, a)E_{a1} \quad \dots\dots\dots\dots\dots\dots \quad (34)$$

$$S^2 \, (\check{E}_{a2}) = (1, a, a^2)\check{E}_{a2} \quad \dots\dots\dots\dots\dots\dots \quad (35)$$

The system of vectors $\check{E}_a$, $\check{E}_b$ and $\check{E}_c$ in this notation is represented as follows:

$$S \, (\check{E}_a) = S^0 \, (\check{E}_{a0}) + S^1 \, (\check{E}_{a1}) + S^2 \, (\check{E}_{a2}) \quad \dots\dots\dots \quad (36)$$

These operators possess interesting properties which when taken advantage of simplify analysis considerably.

Addition and Subtraction.—Two vectors operated on individually by a given sequence operator are equal to the sum of the vectors operated on by the sequence operator. For example:

$$\left.\begin{aligned}
S^1 \, (\check{E}_a) + S^1 \, (\check{E}_b) &= S^1 \, (\check{E}_a + \check{E}_b) \\
S^0 \, (\check{E}_a) + S^0 \, (\check{E}_b) &= S^0 \, (\check{E}_a + \check{E}_b) \\
S^1 \, (\check{E}_a) - S^1 \, (\check{E}_b) &= S^1 \, (\check{E}_a - \check{E}_b)
\end{aligned}\right\} \dots\dots\dots\dots\dots (37)$$

These relations are more or less evident. It should be noted, however, that the operators must be of the same phase sequence.

Transfer of Vector Multiplier.—A vector multiplier may be transferred from outside the vector operator to the inside.

$$\check{K} \, S^1 \, (\check{E}) = S^1 \, (\check{K} \, \check{E}) , \dots\dots\dots\dots\dots\dots\dots \quad (38)$$

Multiplication of Sequence Operators.— The exponents of the sequence operators follow the law of the exponents of a in multiplication. ($a \times a = a^2 ; a^2 \times a = a^3 = 1$).

$$S^1 \, (\check{E}) \times S^1 \, (\check{I}) = S^2 \, (\check{E} \, \check{I})$$

Before considering this statement it is necessary to state just what is meant by this type of multiplication. $S^1 \, (\check{E})$ and $S^1 \, (\check{I})$ each consist of three terms. The multiplication will be assumed to occur symmetrically, the first terms, the second terms and third terms of the groups being multiplied respectively. There is no cross multiplication because each term represents a separate vector. Then,

$$S^1 \, (\check{E}) \times S^1 \, (\check{I}) = (\check{E}, a^2 \, \check{E}, a \, \check{E}) \times \check{I}, a^2 \, \check{I}, a \, \check{I})$$

$$\check{E} \text{ multiplied by } \check{I} = \check{E} \, \check{I}$$

$$a^2 \, \check{E} \text{ multiplied by } a^2 \, \check{I} = a^4 \, \check{E} \, \check{I} = a \, \check{E} \, \check{I}$$

$$a \, \check{E} \text{ multiplied by } a \, \check{I} = a^2 \, \check{E} \, \check{I}$$

So that the product is,

$$(\check{E} \, \check{I}, a \, \check{E} \, \check{I}, a^2 \, (\check{E} \, \check{I})$$

which is equal to,

$$(1, a, a^2) \, \check{E} \, \check{I} = S^2 \, (\check{E} \, \check{I})$$

Similarly,

$$S^2 \, (\check{E}) \times S^2 \, (\check{I}) = (1, a, a^2) \, \check{E} \times (1, a, a^2) \, \check{I}$$

$$= (1, a^2, a) \, \check{E} \, \check{I}$$

$$= S^1 \, (\check{E} \, \check{I})$$

It is evident that adding the exponents (2 + 2) produces 4. This result is similar to that produced by $a^2 \times a^2 = a^4 = a$. The same law also prevails in the application to the sequence operators. This may be further illustrated by,

$$S^2 (\overset{\vee}{E}) \times S^1 (\overset{\vee}{I}) = (1, a^2, a) \overset{\vee}{E} \times (1, a, a^2) \overset{\vee}{I}$$
$$= (1, 1, 1) \overset{\vee}{E} \overset{\vee}{I}$$
$$= S^0 (\overset{\vee}{E} \overset{\vee}{I})$$

This relation may be generalized into,

$$S^m (\overset{\vee}{E}) \times S^n (\overset{\vee}{I}) = S^{m+n} (\overset{\vee}{E} \overset{\vee}{I}) \dots\dots\dots\dots\dots\dots (39)$$

Just as $\dfrac{d}{dt}$, $\sqrt{}$, sin, etc., by themselves mean nothing and only

take on significance when followed by a function of some kind, so S^0, S^1, and S^2 by themselves mean nothing. When it is desired to express the three vectors $1, a^2, a$, the operator must be followed by unity, e.g. $S^1 (1)$.

CHANGE OF FIRST TERM IN SEQUENCE

In all of the discussion so far phase A has been used as the reference phase the first term in the sequence. The notation

$$S(\overset{\vee}{E}_a) = S^0 \overset{\vee}{E}_{a0} + S^1 \overset{\vee}{E}_{a1} + S^2 \overset{\vee}{E}_{a2}$$

means that $\overset{\vee}{E}_a$ is the first term in the sequence. It sometimes becomes desirable to change the first term. Instead of considering the vectors $\overset{\vee}{E}_a$, $\overset{\vee}{E}_b$, $\overset{\vee}{E}_c$, one might wish to consider them in the order $\overset{\vee}{E}_b$, $\overset{\vee}{E}_c$, $\overset{\vee}{E}_a$. This can be accomplished as follows:

Using phase A as datum:
$$\overset{\vee}{E}_{a0} = 1/3 (\overset{\vee}{E}_a + \overset{\vee}{E}_b + \overset{\vee}{E}_c)$$
$$\overset{\vee}{E}_{a1} = 1/3 (\overset{\vee}{E}_a + a \overset{\vee}{E}_b + a^2 \overset{\vee}{E}_c)$$
$$\overset{\vee}{E}_{a2} = 1/3 (\overset{\vee}{E}_a + a^2 \overset{\vee}{E}_b + a \overset{\vee}{E}_c)$$

Using phase B as datum:

$$\check{E}_{b0} = 1/3 \, (\check{E}_b + \check{E}_c + \check{E}_a) = \check{E}_{a0}$$
$$\check{E}_{b1} = 1/3 \, (\check{E}_b + a \check{E}_c + a^2 \check{E}_a)$$
$$\qquad = a^2 \, 1/3 \, (a \check{E}_b + a^2 \check{E}_c + \check{E}_a) = a^2 \check{E}_{a1} \Bigg\} \quad \ldots \ldots \quad (40)$$
$$\check{E}_{b2} = 1/3 \, (\check{E}_b + a^2 \check{E}_c + a \check{E}_a)$$
$$\qquad = a \, 1/3 \, (a^2 \check{E}_b + a \check{E}_c + \check{E}_a) = a \check{E}_{a2}$$

Using phase C as datum:

$$\check{E}_{c0} = 1/3 \, (\check{E}_c + \check{E}_a + \check{E}_b) = \check{E}_{a0}$$
$$\check{E}_{c1} = 1/3 \, (\check{E}_c + a \check{E}_a + a^2 \check{E}_b)$$
$$\qquad = 1/3 \, (a^2 \check{E}_c + \check{E}_a + a \check{E}_b) = a \check{E}_{a1} \Bigg\} \quad \ldots \ldots \quad (41)$$
$$\check{E}_{c2} = 1/3 \, (\check{E}_c + a^2 \check{E}_a + a \check{E}_b)$$
$$\qquad = a^2 \, 1/3 \, (a \check{E}_c + \check{E}_a + a^2 \check{E}_b) = a^2 \check{E}_{a2}$$

Now since

$$S \, (\check{E}_b) = S^0 \, \check{E}_{b0} + S^1 \, \check{E}_{b1} + S^2 \, \check{E}_{b2} \quad \ldots \ldots \ldots \ldots \quad (42)$$

and

$$S \, (\check{E}_c) = S^0 \, \check{E}_{c0} + S^1 \, \check{E}_{c1} + S^2 \, \check{E}_{c2} \quad \ldots \ldots \ldots \ldots \quad (43)$$

Substituting the above values of components, we obtain:

$$S \, (\check{E}_b) = S^0 \, \check{E}_{a0} + S^1 \, a^2 \check{E}_{a1} + S^2 \, a \check{E}_{a2} \quad \ldots \ldots \ldots \quad (44)$$
$$S \, (\check{E}_c) = S^0 \, \check{E}_{a0} + S^1 \, a \check{E}_{a1} + S^2 \, a^2 \check{E}_{a2} \quad \ldots \ldots \ldots \quad (45)$$

Equation (42) expresses the three vectors $\check{E}_b$, $\check{E}_c$, and $\check{E}_a$ in terms of the components, using $\check{E}_b$, as the first term or datum, while equation (44) expresses the same vectors in terms of the sequence components using $\check{E}_a$ as the first term. Similarly equations (43) and (45) express the three vectors $\check{E}_c$, $\check{E}_a$, $\check{E}_b$, in terms of their sequence components using $\check{E}_c$ and $\check{E}_a$, respectively as the first term.

STAR-DELTA TRANSFORMATIONS

Problems sometimes arise involving both star and delta currents and voltages. The conditions are such that the equations are set up more easily in terms of both star and delta quantities. It is convenient for the solution of such problems to be able to convert readily from star to delta and vice-versa in terms of the symmetrical components.

Let $\check{I}_A$, $\check{I}_B$, and $\check{I}_C$, be the delta currents in the delta connected windings of a machine and $\check{I}_a$, $\check{I}_b$, and $\check{I}_c$, the currents in the line in Fig. B-6. In terms of their sequence components the two systems may be expressed as follows:

$$S\,(\check{I}_A) = S^0\,\check{I}_{A\,0} + S^1\,\check{I}_{A\,1} + S^2\,\check{I}_{A\,2} \;\dots\dots\dots\dots (46)$$

$$S\,(\check{I}_a) = S^0\,\check{I}_{a0} + S^1\,\check{I}_{a1} + S^2\,\check{I}_{a2} \;\dots\dots\dots\dots (47)$$

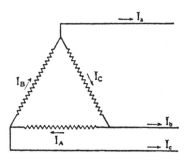

Fig. B-6. Currents in a Delta-Connected System

At the junctions the following relations are satisfied:

$$\check{I}_a = \check{I}_B - \check{I}_C$$
$$\check{I}_b = \check{I}_C - \check{I}_A$$
$$\check{I}_c = \check{I}_A - \check{I}_B$$

or

$$S\,(\check{I}_a) = S\,(\check{I}_B) - S\,(\check{I}_C) \;\dots\dots\dots\dots\dots (48)$$

Using the relations from equations (46) and (47) the first terms in the last two expressions may be changed, so that on substituting the values of $S(\check{I}_A)$ and $S(\check{I}_a)$ from equations (46) and (47), equation (48) becomes:

$$S^0 \, \check{I}_{a0} + S^1 \, \check{I}_{a1} + S^2 \, \check{I}_{a2}$$
$$= S^0 \, \check{I}_{A0} + S^1 \, a^2 \, \check{I}_{A1} + S^2 \, a \, \check{I}_{A2}$$
$$- S^0 \, \check{I}_{A0} - S^1 \, a \, \check{I}_{A1} - S^2 \, a^2 \, \check{I}_{A2}$$

From which

$$S^0 \, \check{I}_{a0} = S^0 \, (\check{I}_{A0} - \check{I}_{A0}) = 0$$
$$S^1 \, \check{I}_{a1} = S^1 \, (a^2 - a) \, \check{I}_{A1} = S^1 \, (-j\sqrt{3})\check{I}_{A1}$$
$$S^2 \, \check{I}_{a2} = S^2 \, (a - a^2) \, \check{I}_{A2} = S^2 \, (+j\sqrt{3})\check{I}_{A2}$$

or

$$\left. \begin{array}{ll}
\check{I}_{a0} = 0 & \check{I}_{A0} = \text{indeterminate} \\[2ex]
\check{I}_{a1} = -j\sqrt{3} \, \check{I}_{A1} & \check{I}_{A1} = \dfrac{+j}{\sqrt{3}}\check{I}_{a1} \\[2ex]
\check{I}_{a2} = +j\sqrt{3} \, \check{I}_{A2} & \check{I}_{A2} = \dfrac{-j}{\sqrt{3}}\check{I}_{a2}
\end{array} \right\} \quad \cdots \cdots \cdots (49)$$

These relations are indicated in the vector diagram, Fig. B-7.

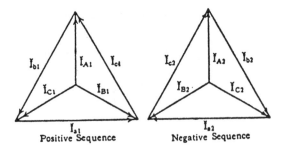

Fig. B-7. Current Relationships

It is apparent that no zero sequence current can exist in lines feeding a delta connection, and that currents of zero sequence circulating in the delta are indeterminant from line currents alone.

Now let $\check{E}_A$, $\check{E}_B$, and $\check{E}_C$, (Fig. B-8), be the delta voltages and $\check{E}_a$ $\check{E}_b$, and $\check{E}_c$, the star voltages. What are the corresponding equivalences as obtained for currents?

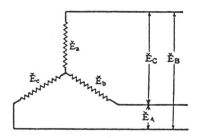

Fig. B-8. Voltages in a Star-Connected System

In terms of their sequence components the two systems may be expressed as follows:

$$S\,(\check{E}_a) = S^0\,\check{E}_{a0} + S^1\,\check{E}_{a1} + S^2\,\check{E}_{a2} \quad \dots\dots\dots\dots (50)$$

$$S\,(\check{E}_A) = S^0\,\check{E}_{A0} + S^1\,\check{E}_{A1} + S^2\,\check{E}_{A2} \quad \dots\dots\dots\dots (51)$$

By definition let

$$\left.\begin{aligned}
\check{E}_A &= \check{E}_c - \check{E}_b \\
\check{E}_B &= \check{E}_a - \check{E}_c \\
\check{E}_C &= \check{E}_b - \check{E}_a
\end{aligned}\right\} \dots\dots\dots\dots\dots\dots\dots (51a)$$

(With equal justification we could have assumed $\check{E}_A = \check{E}_b - \check{E}_c$, etc. The choice is purely arbitrary.)

or

$$S\,(\check{E}_A) = S\,(\check{E}_c) - S\,(\check{E}_b) \quad \dots\dots\dots\dots\dots (52)$$

Changing the first term in the last two expressions and sub-stituting we obtain:

$$S^0 \check{E}_{A0} + S^1 \check{E}_{A1} + S^2 \check{E}_{A2} = S^0 \check{E}_{a0} + S^1 a \check{E}_{a1}$$
$$+ S^2 a^2 \check{E}_{a2} - S^0 \check{E}_{a0} - S^1 a^2 \check{E}_{a1} - S^2 a \check{E}_{a2}$$

From which

$$S^0 \check{E}_{A0} = S^0 (\check{E}_{a0} - \check{E}_{a0}) = 0$$
$$S^1 \check{E}_{A1} = S^1 (a - a^2) \check{E}_{a1} = S^1 (+ j\sqrt{3}) \check{E}_{a1}$$
$$S^2 \check{E}_{A2} = S^2 (a^2 - a) \check{E}_{a2} = S^2 (- j\sqrt{3}) \check{E}_{a2}$$

$$\left.\begin{array}{ll} \check{E}_{a0} = \text{indeterminate} \quad \check{E}_{A0} = 0 \\[2mm] \check{E}_{a1} = - \dfrac{j}{\sqrt{3}} \check{E}_{A1} \qquad \check{E}_{A1} = (j\sqrt{3}) \check{E}_{a1} \\[4mm] \check{E}_{a2} = + \dfrac{j}{\sqrt{3}} \check{E}_{A2} \qquad \check{E}_{A2} = (-j\sqrt{3}) \check{E}_{a2} \end{array}\right\} \dots\dots (53)$$

These relations are indicated in the vector diagram in Fig. B-9. Because the sum of the delta voltage must always form a closed triangle, i.e., add to zero, the delta voltages can contain no zero sequence quantity.

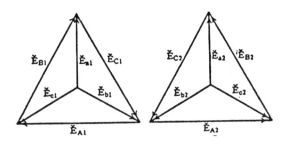

Fig. B-9. Voltage Relationship

II. RESOLUTION OF THREE VECTOR PRODUCTS INTO PRODUCTS OF THEIR SYMMETRICAL COMPONENTS

The previous article discussed some of the fundamental theory of the method of symmetrical components, describing the method for the resolution of voltage or current vectors of a three-phase system into their symmetrical components. The present article concludes the general discussion of the fundamental theory and is concerned chiefly with the resolution of voltage drops into their symmetrical components. This article develops the great simplification that obtains when the system is symmetrical at all points except the point at which a single-phase load or fault is applied.

The determination of voltage drops produced by currents flowing through unsymmetrical impedances will now be considered. Assume three impedances, $\overset{v}{Z}_a$, $\overset{v}{Z}_b$, and $\overset{v}{Z}_c$ connected in star to a neutral wire as shown in Fig. B-10. Let currents $\overset{v}{I}_a$, $\overset{v}{I}_b$, and $\overset{v}{I}_c$ flow in the respective phases. Considering the individual phases the equations expressing the relations between currents and voltages are:

$$\overset{v}{E}_a = \overset{v}{Z}_a \overset{v}{I}_a \dots\dots\dots\dots\dots\dots\dots\dots\dots (54)$$

$$\overset{v}{E}_b = \overset{v}{Z}_b \overset{v}{I}_b \dots\dots\dots\dots\dots\dots\dots\dots\dots (55)$$

$$\overset{v}{E}_c = \overset{v}{Z}_c \overset{v}{I}_c \dots\dots\dots\dots\dots\dots\dots\dots\dots (56)$$

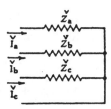

Fig. B-10. Star Connected Impedances

In general, the impedances $\overset{v}{Z}_a$, $\overset{v}{Z}_b$, $\overset{v}{Z}_c$ will be different in magnitude and power-factor. In general, also, the currents may be resolved into the three symmetrical components, the equations for which may be written as follows:

$$\overset{v}{I}_a = \overset{v}{I}_{a1} + \overset{v}{I}_{a2} + \overset{v}{I}_{a0} \quad\dots\dots\dots\dots\dots\dots (57)$$

$$\overset{v}{I}_b = a^2 \overset{v}{I}_{a1} + a \overset{v}{I}_{a2} + \overset{v}{I}_{a0} \quad\dots\dots\dots\dots (58)$$

$$\overset{v}{I}_c = a \overset{v}{I}_{a1} + a^2 \overset{v}{I}_{a2} + \overset{v}{I}_{a0} \quad\dots\dots\dots\dots (59)$$

Substituting the value of currents from equations (57) to (59) in equations (54) to (56) gives:

$$\overset{v}{E}_a = (\overset{v}{Z}_a) \overset{v}{I}_{a1} + (\overset{v}{Z}_a) \overset{v}{I}_{a2} + (\overset{v}{Z}_a) \overset{v}{I}_{a0} \quad\dots\dots\dots\dots (60)$$

$$\overset{v}{E}_b = (a^2 \overset{v}{Z}_b) \overset{v}{I}_{a1} + (a \overset{v}{Z}_b) \overset{v}{I}_{a2} + (\overset{v}{Z}_b) \overset{v}{I}_{a0} \quad\dots\dots (61)$$

$$\overset{v}{E}_c = (a \overset{v}{Z}_c) \overset{v}{I}_{a1} + (a^2 \overset{v}{Z}_c) \overset{v}{I}_{a2} + (\overset{v}{Z}_c) \overset{v}{I}_{a0} \quad\dots\dots (62)$$

As shown in the preceding article the voltages $\overset{v}{E}_a$, $\overset{v}{E}_b$, and $\overset{v}{E}_c$, may be resolved into their symmetrical components as follows:

$$\overset{v}{E}_{a1} = 1/3 (\overset{v}{E}_a + a \overset{v}{E}_b + a^2 \overset{v}{E}_c) \quad\dots\dots\dots\dots (63)$$

$$\overset{v}{E}_{a2} = 1/3 (\overset{v}{E}_a + a^2 \overset{v}{E}_b + a \overset{v}{E}_c) \quad\dots\dots\dots\dots (64)$$

$$\overset{v}{E}_{a0} = 1/3 (\overset{v}{E}_a + \overset{v}{E}_b + \overset{v}{E}_c) \quad\dots\dots\dots\dots\dots (65)$$

The various symmetrical components of voltages may now be expressed in terms of impedance drops by substituting the values of $\overset{v}{E}_a$, $\overset{v}{E}_b$ and $\overset{v}{E}_c$, from equations (60) to (62) in equations (63) to (65) with the following results:

$$\overset{v}{E}_{a1} = 1/3 (\overset{v}{Z}_a + \overset{v}{Z}_b + \overset{v}{Z}_c) \overset{v}{I}_{a1} + 1/3 (\overset{v}{Z}_a + a^2 \overset{v}{Z}_b$$
$$+ a \overset{v}{Z}_c) \overset{v}{I}_{a2} + 1/3 (\overset{v}{Z}_a + a \overset{v}{Z}_b + a^2 \overset{v}{Z}_c) \overset{v}{I}_{a0} \quad\dots (66)$$

$$\overset{v}{E}_{a2} = 1/3 (\overset{v}{Z}_a + a \overset{v}{Z}_b + a^2 \overset{v}{Z}_c) \overset{v}{I}_{a1} + 1/3 (\overset{v}{Z}_a$$
$$+ \overset{v}{Z}_b + \overset{v}{Z}_c) \overset{v}{I}_{a2} + 1/3 (\overset{v}{Z}_a + a^2 \overset{v}{Z}_b + a \overset{v}{Z}_c) \overset{v}{I}_{a0} \quad\dots (67)$$

$$\overset{v}{E}_{a0} = 1/3(\overset{v}{Z}_a + a^2\,\overset{v}{Z}_b + a\overset{v}{Z}_c)\,\overset{v}{I}_{a1} + 1/3(\overset{v}{Z}_a + a\,\overset{v}{Z}_b$$
$$+ a^2\,\overset{v}{Z}_c)\overset{v}{I}_{a2} + 1/3(\overset{v}{Z}_a + \overset{v}{Z}_b + Z_c)\overset{v}{I}_{a0} \cdots \cdots (68)$$

It will be observed that the voltage drop produced by the circulation of current through unsymmetrical impedances is made up of three impedance drops. Closer examination will show that while nine terms are involved there are only three distinct impedances:

$$1/3\,(\overset{v}{Z}_a + \overset{v}{Z}_b + \overset{v}{Z}_c),\; 1/3\,(\overset{v}{Z}_a + a\,\overset{v}{Z}_b + a^2\,\overset{v}{Z}_c),$$
$$1/3\,(\overset{v}{Z}_a + a^2\,\overset{v}{Z}_b + a\,\overset{v}{Z}_c)$$

These terms are impedances since they are the ratio of the voltages impressed to the currents flowing. It will be noted that the voltage drops may be arranged to form three symmetrical terms forming a positive, a negative and a zero sequence system. Study of the voltage drops will show that the voltage drops of one symmetrical component are not always produced by currents of the corresponding symmetrical component. There is, moreover, a certain sequence which may be observed in the arrangement of terms producing the voltage drops. This makes possible an important simplification in the general theory of the method of symmetrical components. This is obtained by forming the voltage drops with the use of the sequence operator.

The relations which have been developed in the preceding paragraphs can be stated more generally in the following theorem:

Theorem:—Any three vector products $S(\overset{v}{K}_x\overset{v}{I}_a) =$

$$\overset{v}{K}_x\overset{v}{I}_a,\; \overset{v}{K}_y\overset{v}{I}_b,\; \overset{v}{K}_z\overset{v}{I}_c = S^0\,(\overset{v}{K}_0\overset{v}{I}_0 + \overset{v}{K}_2\overset{v}{I}_1 + \overset{v}{K}_1\overset{v}{I}_2)$$
$$+ S^1\,(\overset{v}{K}_1\overset{v}{I}_0 + \overset{v}{K}_0\overset{v}{I}_1 + \overset{v}{K}_2\overset{v}{I}_2) + S^2\,(\overset{v}{K}_2\overset{v}{I}_0$$
$$+ \overset{v}{K}_1\overset{v}{I}_1 + \overset{v}{K}_0\overset{v}{I}_2) \cdots \cdots \cdots (69)$$

where

$$\check{K}_0 = 1/3\,(\check{K}_x + \check{K}_y + \check{K}_z)$$
$$\check{K}_1 = 1/3\,(\check{K}_x + a\,\check{K}_y + a^2\,\check{K}_z)$$
$$\check{K}_2 = 1/3\,(\check{K}_x + a^2\,\check{K}_y + a\,\check{K}_z)$$
$$\check{I}_0 = 1/3\,(\check{I}_a + \check{I}_b + \check{I}_c)$$
$$\check{I}_1 = 1/3\,(\check{I}_a + a\,\check{I}_b + a^2\,\check{I}_c)$$
$$\check{I}_2 = 1/3\,(\check{I}_a + a^2\,\check{I}_b + a\,\check{I}_c)$$

$$\left.\right\} \quad \dots\dots\dots\dots\dots\dots (70)$$

The proof of this relation may be demonstrated as follows: Substituting the values of the individual terms in the original products gives:

$$\check{K}_x\check{I}_a = (\check{K}_0 + \check{K}_1 + \check{K}_2)(\check{I}_0 + \check{I}_1 + \check{I}_2)$$
$$\check{K}_y\check{I}_b = (\check{K}_0 + a^2\,\check{K}_1 + a\,\check{K}_2)(\check{I}_0 + a^2\,\check{I}_1 + a\,\check{I}_2)$$
$$\check{K}_z\check{I}_c = (\check{K}_0 + a\,\check{K}_1 + a^2\,\check{K}_2)(\check{I}_0 + a\,\check{I}_1 + a^2\,\check{I}_2)$$

Expanding these expressions and rearranging:

$$\check{K}_x\check{I}_a = (\check{K}_0\check{I}_0 + \check{K}_2\check{I}_1 + \check{K}_1\check{I}_2) + \check{K}_1\check{I}_0 + \check{K}_0\check{I}_1$$
$$+ \check{K}_2\check{I}_2) + (\check{K}_2\check{I}_0 + \check{K}_1\check{I}_1 + \check{K}_0\check{I}_2)$$
$$\check{K}_y\check{I}_b = (\check{K}_0\check{I}_0 + \check{K}_2\check{I}_1 + \check{K}_1\check{I}_2) + a^2\,(\check{K}_1\check{I}_0$$
$$+ \check{K}_0\check{I}_1 + \check{K}_2\check{I}_2) + a(\check{K}_2\check{I}_0 + \check{K}_1\check{I}_1 + \check{K}_0\check{I}_2)$$
$$\check{K}_z\check{I}_c = (\check{K}_0\check{I}_0 + \check{K}_2\check{I}_1 + \check{K}_1\check{I}_2) + a\,(\check{K}_1\check{I}_0 + \check{K}_0\check{I}_1$$
$$+ \check{K}_2\check{I}_2) + a^2\,(\check{K}_2\check{I}_0 + \check{K}_1\check{I}_1 + \check{K}_0\check{I}_2)$$

$$\left.\right\} \quad (71)$$

which proves the above theorem.

An interesting fact concerning these expressions is that the first, second and third terms in parentheses in each equation are equal. In addition the coefficients of the first term form the zero sequence, those of the second term the positive sequence, and those of the third term the negative sequence. This suggests the following short-cut notation to represent the relation:

$$S(\breve{K}_z \breve{I}_a) = \left\{ \begin{array}{c} \breve{K}_x \breve{I}_a \\ \breve{K}_y \breve{I}_b \\ \breve{K}_z \breve{I}_c \end{array} \right\} = S^0 \, (\breve{K}_0 \breve{I}_0 + \breve{K}_2 \breve{I}_1 + \breve{K}_1 \breve{I}_2)$$

$$+ S^1 (\breve{K}_1 \breve{I}_0 + \breve{K}_2 \breve{I}_2) + S^2 (\breve{K}_2 \breve{I}_0$$

$$+ \breve{K}_1 \breve{I}_1 + \breve{K}_0 \breve{I}_2) \, \ldots\ldots\ldots\ldots\ldots\ldots\ldots\ldots \quad (72)$$

Using the laws developed for the sequence operators these
same results can be obtained by a much more elegant method:

$$\breve{K}_x \breve{I}_a, \, \breve{K}_y \breve{I}_b, \, \breve{K}_z \breve{I}_c = S(\breve{K}_z \breve{I}_a)$$

$$= S(\breve{K}_x) \, S(\breve{I}_a)$$

$$= [S^0(\breve{K}_0) + S^1(\breve{K}_1) + S^2(\breve{K}_2)] \, [S^0(\breve{I}_0)$$

$$+ S^1(\breve{I}_1) + S^2(\breve{I}_2)]$$

$$= S^0(\breve{K}_0) \, S^0(\breve{I}_0) + S^0(\breve{K}_0) \, S^1(\breve{I}_1) + S^0(\breve{K}_0) \, S^2(\breve{I}_2)$$

$$+ S^1(\breve{K}_1) \, S^0(\breve{I}_0) + S^1(\breve{K}_1) \, S^1(\breve{I}_1) + S^1(\breve{K}_1) \, S^2(\breve{I}_2)$$

$$+ S^2(\breve{K}_2) \, S^0(\breve{I}_0) + S^2(\breve{K}_2) \, S^1(\breve{I}_1) + S^2(\breve{K}_2) \, S^2(\breve{I}_2)$$

Recalling the laws of exponents as applying to the sequence
operators:

$$S(\breve{K}_x \breve{I}_a) = S^0 \, [\breve{K}_0 \breve{I}_0 + \breve{K}_1 \breve{I}_2 + \breve{K}_2 \breve{I}_1]$$

$$+ S^1 \, [\breve{K}_0 \breve{I}_1 + \breve{K}_1 \breve{I}_0 + \breve{K}_2 \breve{I}_2]$$

$$+ S^2 \, [\breve{K}_0 \breve{I}_2 + \breve{K}_1 \breve{I}_1 + \breve{K}_2 \breve{I}_0]$$

Inasmuch as the value of the subscript always corresponds to
the sequence, the same law also applies to them. The zero
sequence component of the products is comprised of all those
terms whose subscripts can add to zero (or multiples of 3), the
positive sequence components of those terms which can add to
unity (or 4), and the negative sequence of those which can add
to 2 (or 5).

What of the physical significance of these expressions? This
will be made clear by applying the deductions to the case first

considered, viz., an impedance $\check{Z}_a$ in phase a, $\check{Z}_b$ in phase b, and $\check{Z}_c$ in phase c. The question arises "What are the voltage drops in terms of the symmetrical components of the impedances and currents?" The actual drops may be written

$$S(\check{Z}_a \check{I}_a) = \begin{Bmatrix} \check{Z}_a \check{I}_a \\ \check{Z}_b \check{I}_b \\ \check{Z}_c \check{I}_c \end{Bmatrix} = + S^0 (\check{Z}_0 \check{I}_0 + \check{Z}_2 \check{I}_1 + \check{Z}_1 \check{I}_2)$$

$$+ S^1 (\check{Z}_1 \check{I}_0 + \check{Z}_0 \check{I}_1 + \check{Z}_2 \check{I}_2) + S^2 (\check{Z}_2 \check{I}_0 + \check{Z}_1 \check{I}_1$$

$$+ \check{Z}_0 \check{I}_2) \quad\quad\quad\quad\quad\quad\quad\quad\quad\quad\quad\quad\quad (73)$$

Consider the case in which the three impedances are equal. It is evident for this case that if only currents of zero phase sequence flow, only voltage drops of zero sequence are present. This result may be checked mathematically. When the impedances are equal, $\check{Z}_1$ and $\check{Z}_2$ are equal to zero and since $\check{I}_1$ and $\check{I}_2$ are zero all the terms but $S^0 (\check{Z}_0 \check{I}_0)$ disappear. If only positive sequence currents are present all terms but $S^1 (\check{Z}_0 \check{I}_1)$ disappear and similarly, if only negative sequence currents are present all terms but $S^2 (\check{Z}_0 \check{I}_2)$ disappear.

Consider next the case in which the impedances in the three phases form a positive phase sequence.

$$\left. \begin{array}{l} \check{Z}_a = \check{Z} \\ \check{Z}_b = a^2 \check{Z} \\ \check{Z}_c = a\check{Z} \end{array} \right\} \quad\quad\quad\quad\quad\quad\quad\quad\quad\quad\quad\quad (74)$$

This requires in the impedances both positive and negative reactances and resistances. Negative reactance may be represented by condensers, but how may negative resistances be obtained? Negative resistance characteristics may be obtained by vacuum tube set-ups or by series commutator machines in the leads. Of course, this assumption is only hypothetical. It is merely introduced to clear up the conceptions of unbalanced impedances. In this case

$$\overset{v}{Z_0} = 1/3[\overset{v}{Z} + a^2\overset{v}{Z} + a\overset{v}{Z}] = 0$$
$$\overset{v}{Z_1} = 1/3[\overset{v}{Z} + a(a^2\overset{v}{Z}) + a^2(a\overset{v}{Z})] = \overset{v}{Z} \Big\} \quad \dots\dots\dots\dots (75)$$
$$\overset{v}{Z_2} = 1/3[\overset{v}{Z} + a^2(a^2\overset{v}{Z}) + a(a\overset{v}{Z})] = 0$$

Again assume that only zero sequence currents flow, the drops in this case are $\overset{v}{Z}\overset{v}{I_0}$, $a^2 \overset{v}{Z} \overset{v}{I_0}$, and $a \overset{v}{Z} \overset{v}{I_0}$. It can be seen that these drops form a positive phase sequence; a zero phase sequence current passing through positive phase sequence impedances produces positive phase sequence voltage drops. This conclusion can be verified mathematically by inserting the values of $\overset{v}{Z_0}$, $\overset{v}{Z_1}$, and $\overset{v}{Z_2}$ from equations (75) in equation (73). When only zero sequence currents are present all of the terms except $S^2(\overset{v}{Z_1}\overset{v}{I_0})$ vanish. With only positive sequence current present the drops

$$\overset{v}{Z_a}\overset{v}{I_a}, \overset{v}{Z_b}\overset{v}{I_b}, \overset{v}{Z_c}\overset{v}{I_c} = \overset{v}{Z}\overset{v}{I}, a^2\overset{v}{Z} \cdot a^2\overset{v}{I}, a\overset{v}{Z} \cdot a\overset{v}{I}$$
$$= \overset{v}{Z}\overset{v}{I}, a\,\overset{v}{Z}\overset{v}{I}, a^2\,\overset{v}{Z}\overset{v}{I}$$

are obtained. These drops it will be noted are of negative sequence. Again referring to equation (73) this result will be confirmed as all drops but the $\overset{v}{Z_1}\overset{v}{I_1}$ of the negative sequence vanish.

Similar conceptions may be shown for impedances of negative sequence.

Now comparing the expression for the voltage drops $S(\overset{v}{Z_a}\overset{v}{I_a})$ with the expression for the voltage drop in a single phase circuit $\overset{v}{Z}\overset{v}{I}$, the advantage of the notation by the sequence operators becomes apparent. It enables one to express the same relations for a polyphase system as the simple quantity $\overset{v}{Z}\overset{v}{I}$ expresses for the single-phase system. It simplifies the polyphase system to the same extent as the j operator simplified the single-phase alternating-current system.

UNSYMMETRICAL NETWORK
CONTAINING SELF AND
MUTUAL IMPEDANCES

In very few polyphase circuits are the phases independent of each other; they are usually coupled electromagnetically. In transmission lines the mutual coupling is quite large, but it is usually cancelled out by the method of calculating the self inductance. Only that space field between the conductors is considered. This conveys the impression that the mutual inductance is nil. In polyphase machines such as generators, motors (synchronous and asynchronous), synchronous condensers and induction regulators, the coupling between phases is considerable. The exceptional cases in which the phases are not coupled are lighting and electric furnace loads. In the remainder of this article it will be shown that for circuits in which the impedances are symmetrical, even though they contain mutual inductance, the three-phase sequences may be treated entirely independently; there is no interference between different sequences. This results in considerable simplification; and when it is recalled that with but few exceptions all of our circuits are symmetrical this result becomes extremely important. It will be shown that for such circuits the important constants are the impedances to positive sequence, negative sequence and zero sequence currents.

Those readers who are not particularly interested in the proofs for all the theorems, but are willing to accept the above statements on faith, might omit the remainder of this article with little loss of continuity.

Consider the circuit shown in Fig. B-11. It is desired to express the drop across the impedances in terms of the symmetrical components of the current and self and mutual impedances. Let the current $\dot{I}_a$, $\dot{I}_b$, and $\dot{I}_c$ flow in the three phases respectively, and let the self impedances be $\dot{Z}_{aa}$, $\dot{Z}_{bb}$, and $\dot{Z}_{cc}$. In addition, let $\overset{v}{Z}_{ab}$ be the mutual impedance between phases and b, $\overset{v}{Z}_{bc}$ that between phases b and c, and $\overset{v}{Z}_{ca}$ that between phases c and a.

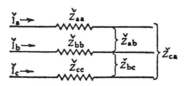

Fig. B-11. Series Impedance Circuit

The drops involved may be arranged in three groups:

I — Potential drop in a due to current in a $= \overset{v}{Z}_{aa}\overset{v}{I}_a$

Potential drop in b due to current in b $= \overset{v}{Z}_{bb}\overset{v}{I}_b$

Potential drop in c due to current in c $= \overset{v}{Z}_{cc}\overset{v}{I}_c$

II — Potential drop in a due to current in b $= \overset{v}{Z}_{ab}\overset{v}{I}_b$

Potential drop in b due to current in c $= \overset{v}{Z}_{bc}\overset{v}{I}_c$

Potential drop in c due to current in a $= \overset{v}{Z}_{ca}\overset{v}{I}_a$

III — Potential drop in a due to current in c $= \overset{v}{Z}_{ac}\overset{v}{I}_c$

Potential drop in b due to current in a $= \overset{v}{Z}_{ba}\overset{v}{I}_a$

Potential drop in c due to current in b $= \overset{v}{Z}_{cb}\overset{v}{I}_b$

The terms of group I represent the voltage drops due to the self impedances and is the same as that just treated. It consists of the three vector products

$$\overset{v}{Z}_{aa}\overset{v}{I}_a, \quad \overset{v}{Z}_{bb}\overset{v}{I}_b, \quad \overset{v}{Z}_{cc}\overset{v}{I}_c$$

Referring to the theorem relating to resolution of vector products into symmetrical form these drops may be expressed as follows:

$$\left.\begin{array}{c} \overset{v}{Z}_{aa}\overset{v}{I}_a \\ \overset{v}{Z}_{bb}\overset{v}{I}_b \\ \overset{v}{Z}_{cc}\overset{v}{I}_c \end{array}\right\} = S(\overset{v}{Z}_{aa}\overset{v}{I}_a)$$

$$\left.\begin{array}{l} = S^0 \ (\overset{v}{Z}_{aa0}\overset{v}{I}_{a0} + \overset{v}{Z}_{aa2}\overset{v}{I}_{a1} + \overset{v}{Z}_{aa1}\overset{v}{I}_{a2} \\ + S^1 \ \overset{v}{Z}_{aa1}\overset{v}{I}_{a0} + \overset{v}{Z}_{aa0}\overset{v}{I}_{a1} + \overset{v}{Z}_{aa2}\overset{v}{I}_{a2}) \\ + S^2 \ (\overset{v}{Z}_{aa2}\overset{v}{I}_{a0} + \overset{v}{Z}_{aa1}\overset{v}{I}_{a1} + \overset{v}{Z}_{aa0}\overset{v}{I}_{a^2}) \end{array}\right\} \dots\dots\dots (76)$$

Where

$$\left.\begin{array}{l} \overset{v}{Z}_{aa0} = 1/3(\overset{v}{Z}_{aa} + \overset{v}{Z}_{bb} + \overset{v}{Z}_{cc}) \\ \overset{v}{Z}_{aa1} = 1/3(\overset{v}{Z}_{aa} + a\,\overset{v}{Z}_{bb} + a^2\,\overset{v}{Z}_{cc}) \\ \overset{v}{Z}_{aa2} = 1/3(\overset{v}{Z}_{aa} + a^2\,\overset{v}{Z}_{bb} + a\,\overset{v}{Z}_{cc}) \end{array}\right\} \dots\dots\dots\dots (77)$$

Considering the terms of Group II, the voltages induced in the three phases a, b, and c are respectively

$$\overset{v}{Z}_{ab}\overset{v}{I}_b, \quad \overset{v}{Z}_{bc}\overset{v}{I}_c, \quad \overset{v}{Z}_{ca}\overset{v}{I}_a$$

Again, using the three product theorem

$$\left.\begin{array}{c} \overset{v}{Z}_{ab}\overset{v}{I}_b \\ \overset{v}{Z}_{bc}\overset{v}{I}_c \\ \overset{v}{Z}_{ca}\overset{v}{I}_a \end{array}\right\} = S \ (\overset{v}{Z}_{ab}\overset{v}{I}_b)$$

$$\left.\begin{array}{l} = S^0 \ (\overset{v}{Z}_{ab0}\overset{v}{I}_{b0} + \overset{v}{Z}_{ab2}\overset{v}{I}_{b1} + \overset{v}{Z}_{ab1}\overset{v}{I}_{b2}) \\ + S^1 \ (\overset{v}{Z}_{ab1}\overset{v}{I}_{b0} + \overset{v}{Z}_{ab0}\overset{v}{I}_{b1} + \overset{v}{Z}_{ab2}\overset{v}{I}_{b2}) \\ + S^2 \ (\overset{v}{Z}_{ab2}\overset{v}{I}_{b0} + \overset{v}{Z}_{ab1}\overset{v}{I}_{b1} + \overset{v}{Z}_{ab0}\overset{v}{I}_{b2}) \end{array}\right\} \dots\dots\dots (78)$$

Where

$$\left.\begin{array}{l} \overset{v}{Z}_{ab0} = 1/3(\overset{v}{Z}_{ab} + \overset{v}{Z}_{bc} + \overset{v}{Z}_{ca}) \\ \overset{v}{Z}_{ab1} = 1/3(\overset{v}{Z}_{ab} + a\,\overset{v}{Z}_{bc} + a^2\,\overset{v}{Z}_{ca}) \\ \overset{v}{Z}_{ab2} = 1/3(\overset{v}{Z}_{ab} + a^2\,\overset{v}{Z}_{bc} + a\,\overset{v}{Z}_{ca}) \end{array}\right\} \dots\dots\dots (79)$$

In the above expressions the nomenclature $\overset{v}{I}_{b1}$ and $\overset{v}{I}_{b2}$ is used to indicate that the first term in the group, the reference or datum vector, has been changed to the b phase. Now, since

$$\overset{v}{I}_{b0} = \overset{v}{I}_{a0}$$

$$\overset{v}{I}_{b1} = a^2 \overset{v}{I}_{a1}$$

$$\overset{v}{I}_{b2} = a \overset{v}{I}_{a2}$$

$$\left. \begin{aligned}
S(\overset{v}{Z}_{ab}\overset{v}{I}_b) = S^0(\overset{v}{Z}_{ab0}\overset{v}{I}_{a0} + \overset{v}{Z}_{ab2}\, a^2 \overset{v}{I}_{a1} + \overset{v}{Z}_{ab1}\, a\, \overset{v}{I}_{a2}) \\
+ S^1(\overset{v}{Z}_{ab1}\overset{v}{I}_{a0} + \overset{v}{Z}_{ab0}\, a^2 \overset{v}{I}_{a1} + \overset{v}{Z}_{ab2} a\, \overset{v}{I}_{a2}) \\
+ S^2(\overset{v}{Z}_{ab2}\overset{v}{I}_{a0} + \overset{v}{Z}_{ab1}\, a^2\, \overset{v}{I}_{a1} + \overset{v}{Z}_{ab0}\, a\overset{v}{I}_{a2})
\end{aligned} \right\} \quad (80)$$

Considering the terms of group III in a similar manner:

$$\left. \begin{aligned}
\overset{v}{Z}_{ac}\overset{v}{I}_c \\
\overset{v}{Z}_{ba}\overset{v}{I}_a \\
\overset{v}{Z}_{cb}\overset{v}{I}_b
\end{aligned} \right\} = S\,(\overset{v}{Z}_{ac}\overset{v}{I}_c)$$

$$\left. \begin{aligned}
= S^0(\overset{v}{Z}_{ac0}\overset{v}{I}_{c0} + \overset{v}{Z}_{ac2}\overset{v}{I}_{c1} + \overset{v}{Z}_{ac1}\overset{v}{I}_{c2}) \\
+ S^1(\overset{v}{Z}_{ac1}\overset{v}{I}_{c0} + \overset{v}{Z}_{ac0}\overset{v}{I}_{c1} + \overset{v}{Z}_{ac2}\overset{v}{I}_{c2}) \\
+ S^2(\overset{v}{Z}_{ac2}\overset{v}{I}_{c0} + \overset{v}{Z}_{ac1}\overset{v}{I}_{c1} + \overset{v}{Z}_{ac0}\overset{v}{I}_{c2})
\end{aligned} \right\} \quad \ldots\ldots\ldots \quad (81)$$

Since

$$\overset{v}{I}_{c0} = \overset{v}{I}_{a0} \qquad\qquad \overset{v}{Z}_{ac} = \overset{v}{Z}_{ac}$$

$$\overset{v}{I}_{c1} = a\, \overset{v}{I}_{a1} \qquad\qquad \overset{v}{Z}_{ba} = \overset{v}{Z}_{ab}$$

$$\overset{v}{I}_{c2} = a^2 \overset{v}{I}_{a2} \qquad\qquad \overset{v}{Z}_{cb} = \overset{v}{Z}_{bc}$$

and

$$\left. \begin{aligned}
\overset{v}{Z}_{ac0} = 1/3(\overset{v}{Z}_{ac} + \overset{v}{Z}_{ba} + \overset{v}{Z}_{cb}) = \overset{v}{Z}_{ab0} \\
\overset{v}{Z}_{ac1} = 1/3(\overset{v}{Z}_{ac} + a\, \overset{v}{Z}_{ba} + a^2 \overset{v}{Z}_{cb}) = a\, \overset{v}{Z}_{ab1} \\
\overset{v}{Z}_{ac2} = 1/3(\overset{v}{Z}_{ac} + a^2 \overset{v}{Z}_{ba} + a\, \overset{v}{Z}_{cb}) = a^2 \overset{v}{Z}_{ab2}
\end{aligned} \right\} \quad \ldots\ldots \quad (82)$$

Therefore

$$
\begin{aligned}
S(\breve{E}_{ac}\breve{I}_c) = S^0(&\breve{Z}_{ab0}\breve{I}_{a0} + \breve{Z}_{ab2}\breve{I}_{a1} + \breve{Z}_{ab1}\breve{I}_{a2}) \\
+ S^1(&a\,\breve{Z}_{ab1}\breve{I}_{a0} + a\,\breve{Z}_{ab0}\breve{I}_{a1} + a\,\breve{Z}_{ab2}\breve{I}_{a2}) \\
+ S^2(&a^2\breve{Z}_{ab2}\breve{I}_{a0} + a^2\breve{Z}_{ab1}\breve{I}_{a1} + a^2\breve{Z}_{ab0}\breve{I}_{a2})
\end{aligned} \tag{83}
$$

Using the a phase as the first term in the sequence, the drop in the three phases, $S(\breve{E}_a)$ is

$$
\begin{aligned}
S(\breve{E}_a) = S^0(&\breve{Z}_{aa0} + 2\,\breve{Z}_{ab0})\breve{I}_{a0} \\
+ S^0(&\breve{Z}_{aa2} + (a^2 + 1)\,\breve{Z}_{ab2})\breve{I}_{a1} \\
+ S^0(&\breve{Z}_{aa1} + (a + 1)\,\breve{Z}_{ab1})\breve{I}_{a2} \\
+ S^1(&\breve{Z}_{aa1} + (1 + a)\,\breve{Z}_{ab1})\breve{I}_{a0} \\
+ S^1(&\breve{Z}_{aa0} + (a^2 + a)\,\breve{Z}_{ab0})\breve{I}_{a1} \\
+ S^1(&\breve{Z}_{aa2} + 2\,a\,\breve{Z}_{ab2})\breve{I}_{a2} \\
+ S^2(&\breve{Z}_{aa2} + (1 + a^2)\,\breve{Z}_{ab2})\breve{I}_{a0} \\
+ S^2(&\breve{Z}_{aa1} + 2\,a^2\breve{Z}_{ab1})\breve{I}_{a1} \\
+ S^2(&\breve{Z}_{aa0} + (a + a^2)\,\breve{Z}_{ab0})\breve{I}_{a2}
\end{aligned} \tag{84}
$$

or

$$
\begin{aligned}
S(\breve{E}_a) = S^0(&\breve{Z}_{aa0} + 2\,\breve{Z}_{ab0})\breve{I}_{a0} \\
+ S^0(&\breve{Z}_{aa2} - a\,\breve{Z}_{ab2})\breve{I}_{a1} \\
+ S^0(&\breve{Z}_{aa1} - a^2\breve{Z}_{ab1})\breve{I}_{a2} \\
+ S^1(&\breve{Z}_{aa1} - a^2\breve{Z}_{ab1})\breve{I}_{a0} \\
+ S^1(&\breve{Z}_{aa0} - \breve{Z}_{ab0})\breve{I}_{a1} \\
+ S^1(&\breve{Z}_{aa2} + 2a\breve{Z}_{ab2})\breve{I}_{a2} \\
+ S^2(&\breve{Z}_{aa2} - a\,\breve{Z}_{ab2})\breve{I}_{a0} \\
+ S^2(&\breve{Z}_{aa1} + 2\,a^2\breve{Z}_{ab1})\breve{I}_{a1} \\
+ S^2(&\breve{Z}_{aa0} - \breve{Z}_{ab0})\breve{I}_{a2}
\end{aligned} \tag{85}
$$

The above equation gives the general solution regardless of the value of the impedances. Consider the special case in which the impedances are perfectly symmetrical.

$$\left. \begin{array}{l} \overset{v}{Z}_{aa} = \overset{v}{Z}_{bb} = \overset{v}{Z}_{cc} \\ \overset{v}{Z}_{ab} = \overset{v}{Z}_{bc} = \overset{v}{Z}_{ac} \end{array} \right\} \quad \dots\dots\dots\dots\dots\dots\dots\dots\dots \quad (86)$$

Substituting in the equations for the symmetrical components of the impedances in equations (77), (79), and (82), the following is obtained.

$$\left. \begin{array}{l} \overset{v}{Z}_{aa0} = \overset{v}{Z}_{aa} \\ \overset{v}{Z}_{ab0} = \overset{v}{Z}_{ab} \\ \overset{v}{Z}_{aa1} = \overset{v}{Z}_{aa2} = \overset{v}{Z}_{ab1} = \overset{v}{Z}_{ab2} = 0 \end{array} \right\} \quad \dots\dots\dots\dots\dots \quad (87)$$

The general equation for voltage, equation (32) then reduces to

$$\left. \begin{array}{l} S(\overset{v}{E}_a) = S^0\,(\overset{v}{Z}_{aa} + 2\,\overset{v}{Z}_{ab})\overset{v}{I}_{a0} \\ \qquad + S^1\,(\overset{v}{Z}_{aa} - \overset{v}{Z}_{ab})\overset{v}{I}_{a1} \\ \qquad + S^2\,(\overset{v}{Z}_{aa} - \overset{v}{Z}_{ab})\overset{v}{I}_{a2} \end{array} \right\} \quad \dots\dots\dots\dots\dots \quad (88)$$

This equation is quite significant. It indicates that zero sequence voltage drops are produced only by zero sequence voltages, positive sequence voltage drops by positive sequence currents and negative sequence voltage drops by negative sequence currents. The three sequences are separate and distinct and do not react one upon the other. One can readily see the importance of such a fundamental relation.

Although this demonstration is limited to lumped self and mutual impedances the above relations may be shown to hold equally well for distributed constants, both inductance and capacitance. A rigorous proof for this condition is too long for presentation in these articles. A critical examination of the proof for the lumped constants would lead one to expect such relation.

To this point no mention has been made of rotating machine impedance; only static networks have been considered. Mr. Fortescue in his classic has shown that even for symmetrical rotating machines the different phase sequences do not interact one upon the other.

Because of their importance in the method of symmetrical components special names have been assigned to the coefficients in equation (88). The coefficient of $\overset{v}{I}_{a0}$ is called the "impedance to zero sequence currents," that of $\overset{v}{I}_{a1}$ is called the "impedance to positive sequence currents," and that of I_{a2} the "impedance to negative sequence currents." For static networks, or machines with either lumped or distributed constants, the impedances to negative and positive sequences are equal; but in rotating machines these impedances are in general different, e.g., in induction motors the impedance to negative sequence currents is very nearly equal to the impedance with blocked rotor and changes very little with normal variations in slip, but the impedance to positive sequence currents varies considerably with normal variations in slip, being very high with low power-factor at no load.

These impedances are of very great practical importance, and may be measured experimentally with ease. The impedance to positive sequence currents of both series impedances and machine impedances are the impedances ordinarily used in calculations of balanced conditions. The impedance to negative sequence currents may be measured by applying voltages of reverse or negative sequence to the network or machine, while the impedance to zero sequence may be measured by connecting the three phases together and applying single-phase voltage to the group in parallel.

III. CALCULATION OF UNBALANCED FAULTS

In the preceding articles the fundamentals of the method of symmetrical components were developed and discussed. Provided only that the impedances of the three phases be symmetrical, i.e. that, viewed from any phase, both the self and mutual impedances are respectively equal, it was shown for any impedance group containing either self or mutual impedances or both that the three sequences of current or voltage may be considered independently—they do not react one upon the other. Such a system contains only the three impedances of zero sequence. A current of a particular phase sequence gives rise to a voltage of the same sequence only. The ratio of this voltage to the current for any phase sequence is termed the impedance of the circuit to that phase sequence. This relation applies not only to static networks, but to rotating machines as well. A succeeding article will be devoted to the determination of the impedance of different types of circuits and machines to the flow of current of the different sequences. A knowledge of these constants is a prerequisite to the calculation of unbalanced faults. For the present it will be assumed that these constants are known, and methods will be developed for the calculation of system faults. Two general cases will be considered: (1) those systems in which resistance and capacity may be neglected and only reactances taken into account, (2) the more general case in which resistance, charging current, etc., are taken into consideration. Both of these cases will be demonstrated by examples.

KIRCHHOFF'S FIRST LAW

The analysis so far pertains to but one branch. The voltage drops of the various sequences are dependent only upon the current and the impedance of the same sequence. What will be the result when a group or network is considered? At each junction, e.g., the junction in Fig. B-12 in which four circuits converge, by Kirchhoff's Law the sum of currents in all the conductors equals zero:

$$\left.\begin{array}{l} \overset{\vee}{I}_a{}' + \overset{\vee}{I}_a{}'' + \overset{\vee}{I}_a{}''' + \overset{\vee}{I}_a{}'''' = 0 \\[4pt] \overset{\vee}{I}_b{}' + \overset{\vee}{I}_b{}'' + \overset{\vee}{I}_b{}''' + \overset{\vee}{I}_b{}'''' = 0 \\[4pt] \overset{\vee}{I}_c{}' + \overset{\vee}{I}_c{}'' + \overset{\vee}{I}_c{}''' + \overset{\vee}{I}_c{}'''' = 0 \end{array}\right\} \dots \dots \dots \dots \quad (89)$$

or expressed in terms of the sequence operator

$$S(\overset{\vee}{I}_a{}') + S(\overset{\vee}{I}_a{}'') + S(\overset{\vee}{I}_a{}''') + S(\overset{\vee}{I}_a{}'''') = S(0) \dots \dots \dots (90)$$

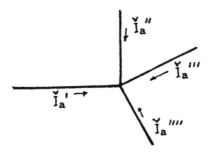

Fig. B-12

Expanded this becomes:

$$S^0(\overset{\vee}{I}_{a0}{}' + \overset{\vee}{I}_{a0}{}'' + \overset{\vee}{I}_{a0}{}''' + \overset{\vee}{I}_{a0}{}'''')$$
$$+ S^1(\overset{\vee}{I}_{a1}{}' + \overset{\vee}{I}_{a1}{}'' + \overset{\vee}{I}_{a1}{}''' + \overset{\vee}{I}_{a1}{}'''')$$
$$+ S^2(\overset{\vee}{I}_{a2}{}' + \overset{\vee}{I}_{a2}{}'' + \overset{\vee}{I}_{a2}{}''' + \overset{\vee}{I}_{a2}{}'''')$$
$$= S^0(0) + S^1(0) + S^2(0)$$

From which

$$\left.\begin{array}{l} S^0(\overset{\vee}{I}_{a0}{}' + \overset{\vee}{I}_{a0}{}'' + \overset{\vee}{I}_{a0}{}''' + \overset{\vee}{I}_{a0}{}'''') = S^0(0) \\[4pt] S^1(\overset{\vee}{I}_{a1}{}' + \overset{\vee}{I}_{a1}{}'' + \overset{\vee}{I}_{a1}{}''' + \overset{\vee}{I}_{a1}{}'''') = S^1(0) \\[4pt] S^2(\overset{\vee}{I}_{a2}{}' + \overset{\vee}{I}_{a2}{}'' + \overset{\vee}{I}_{a2}{}''' + \overset{\vee}{I}_{a2}{}'''') = S^2(0) \end{array}\right\} \dots \dots \dots (91)$$

and

$$
\left.\begin{array}{c}
\overset{v}{I}_{a0}{}' + \overset{v}{I}_{a0}{}'' + \overset{v}{I}_{a0}{}''' + \overset{v}{I}_{a0}{}'''' = 0 \\[4pt]
\overset{v}{I}_{a1}{}' + \overset{v}{I}_{a1}{}'' + \overset{v}{I}_{a1}{}''' + \overset{v}{I}_{a1}{}'''' = 0 \\[4pt]
\overset{v}{I}_{a2}{}' + \overset{v}{I}_{a2}{}'' + \overset{v}{I}_{a2}{}''' + \overset{v}{I}_{a2}{}'''' = 0
\end{array}\right\} \quad \dots\dots\dots\dots\dots \text{(92)}
$$

These equations show that the various sequence components at any junction must individually add up to zero; each sequence of currents obeys Kirchhoff's first law separately. This proof has been given for four circuits, but it is apparent that it is a perfectly general relation which must always hold at any junction and may be extended to any number of conductors.

KIRCHHOFF'S SECOND LAW

Consider a general network of lumped constants, either self or mutual impedances. Since the voltage drops of each phase sequence may be considered separately, the sum of the voltage drops of any sequence around any closed circuit must be equal to the e.m.f.'s of that same sequence. In this respect the network phenomena obey Kirchhoff's second law. This holds true for all points in the system, including the point of fault if this point be considered as a source of zero, negative and positive sequence voltage, $\overset{v}{e}_0$, $\overset{v}{e}_2$, and $\overset{v}{e}_1$, respectively. So, the system may be considered as three separate and distinct networks; one for positive sequence, one for negative sequence and the third for zero sequence. The tie between these systems will be the terminal conditions at the point of fault.

Consider these networks individually. The positive sequence network is in all respects similar to the usual networks considered; the resistances and reactances are the values usually given to calculate line regulation. Each synchronous machine must be considered as a source of e.m.f. which may vary in magnitude and phase position, depending upon the distribution of real and reactive power just previous to the application of the fault. At the point of fault the positive sequence voltage $\overset{v}{e}_1$ will drop, the amount being conditioned upon the type of fault; for three-phase faults it will be zero; for double faults to ground, line to line faults and single-phase faults to ground, it will be higher in the order stated. The exact value is calculable and will be determined.

The negative phase sequence network is in general quite similar to the positive sequence network. Because positive sequence voltages only are generated in the synchronous machines, this network will contain no sources of e.m.f. except the fictitious one at the point of fault.

The zero phase sequence network will likewise be free of internal voltages, the flow of current resulting from the voltage at the point of fault. The impedances to this sequence are radically different from those of either the positive or negative sequences. The line impedances are those obtained by imagining the three conductors connected together, the ground forming the return conductor. The transformer and generator impedances will depend upon the type of connection, whether delta or star connected; if star whether grounded or not. The calculation of the impedances to negative and zero phase sequences will be considered in a subsequent article. For the time being, let it be assumed that these quantities are given.

In particular consider the system shown in Fig. B-13 (a). Two generators A and B at the sending end of a line are connected to two delta-star transformers (only one of which is grounded) which are bussed on the high side. At the receiving end there are two star-star transformer banks (both grounded on high side, but only one on low side) connected to separate generators. The load is assumed concentrated at the receiving end and is partially supplied by the local generators C and D, one of which is grounded.*

---

*The unusual connections shown in this diagram are used merely to illustrate the method as applied to different transformer connections. Line mutual impedances in zero sequence network are neglected.

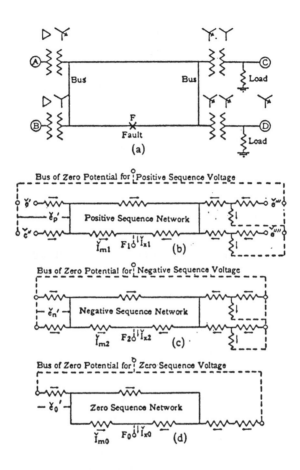

**Fig. B-13. Diagrammatic Representation of the Three Sequence
Networks for a Three-Phase Electrical System**

Assume the fault to occur at the point F midway on one of
the lines. The single line impedance diagram for the positive
sequence is shown in Fig. B-13 (b), and that for the negative
and zero sequence in Fig. B-13 (c) and (d). The synthesis of
these respective networks will be discussed later. Let it suffice
for the time being to point out that the impedances of the
various branches for the different sequences are represented by

the impdenaces in the corresponding topographical position. It
is important to retain the positive flow of current, as indicated
by the arrows, the same in any branch for all three networks,
as the fundamental equations postulate this condition. $\overset{v}{e}{}'$, $\overset{v}{e}{}''$,
$\overset{v}{e}{}'''$, and $\overset{v}{e}{}''''$ are the internal voltages of the four generators A,
B, C, and D respectively. Since positive sequence voltages, only,
are generated within the machines, the corresponding internal
voltages for the negative and zero sequence networks are zero
and, being of the same potential, these points may be connected
together as shown by the dotted lines. Now the only tie between
these otherwise independent networks is the terminal conditions
at the point of fault. The nature of this tie will vary with the
character of the fault. Four cases will be considered in this
analysis:

(1) Ground fault on one conductor.

(2) Double fault to ground.

(3) Single-phase line to line fault.

(4) Three-phase fault.

Imagine three short conductors of zero impedance connected
to the three line conductors at the point of fault. The terminal
conditions imposed by the different types of faults will be applied to these imaginary leads, the potentials to ground of which
will be $\overset{v}{e}_x$, $\overset{v}{e}_y$ and $\overset{v}{e}_z$ respectively, and the currents $\overset{v}{I}_x$, $\overset{v}{I}_y$, and
$\overset{v}{I}_z$. These imaginary conductors are shown in Fig. B-14.

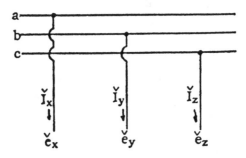

Fig. B-14

(1) *Ground Fault on One Conductor.—* For this case the a phase to be grounded as indicated in Fig. B-15 (a). The terminal conditions in this case are:

$$\left.\begin{array}{l} \overset{v}{e}_x = 0 \\ \overset{v}{I}_y = 0 \\ \overset{v}{I}_z = 0 \end{array}\right\} \dots\dots\dots\dots\dots\dots\dots\dots\dots\dots (93)$$

From equations (30), (31) and (32) of the first article the symmetrical components of current are:

$$\overset{v}{I}_{x0} = 1/3(\overset{v}{I}_x + \overset{v}{I}_y + \overset{v}{I}_z) = 1/3(\overset{v}{I}_x) \dots\dots\dots\dots\dots (94)$$

$$\overset{v}{I}_{x1} = 1/3(\overset{v}{I}_x + a\,\overset{v}{I}_y + a^2\,\overset{v}{I}_z) = 1/3(\overset{v}{I}_x) \dots\dots\dots\dots (95)$$

$$\overset{v}{I}_{x2} = 1/3(\overset{v}{I}_x + a^2\,\overset{v}{I}_y + a\,\overset{v}{I}_z) = 1/3(\overset{v}{I}_x) \dots\dots\dots\dots (96)$$

$$\text{or} \quad \overset{v}{I}_{x0} = \overset{v}{I}_{x2} = \overset{v}{I}_{x1} \dots\dots\dots\dots\dots\dots\dots\dots\dots (97)$$

Also

$$\overset{v}{e}_x = \overset{v}{e}_{x0} + \overset{v}{e}_{x1} + \overset{v}{e}_{x2} = 0 \dots\dots\dots\dots\dots\dots\dots (98)$$

Now from equation (97) it can be seen that the currents at F_1, F_2, and F_0, Fig. B-13, in the three networks for the three sequences are equal, and if the point F_1 of the positive sequence network be connected to the bus of zero potential in the negative sequence network, and the point F_2 in the negative sequence network be connected to the bus of zero potential in the zero sequence network, as shown in Fig. B-15, the three networks will be connected in series and the same total current must flow through the three networks. Since the current distribution within each of the networks is determined solely by the impedances in the individual networks, this combination of the networks enables one to determine the current distribution in any branch. However, for this connection to determine the current completely the relation expressed in equation (98) must also be satisfied.

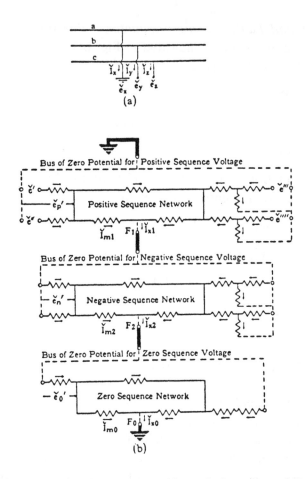

**Fig. B-15. The Three Sequence Networks for a Ground Fault
on One Conductor**

Now starting from any one of the internal voltages of Fig. B-15, say $\overset{v}{e}{}'$, and tracing one's way through the network by any path to F, it can be seen that the positive sequence voltage $\overset{v}{e}_1$ at F is equal to

$$\overset{v}{e}{}' - \Sigma \text{ positive sequence voltage drops from } \overset{v}{e}{}' \text{ to the point F} \quad \dots\dots\dots\dots\dots\dots\dots\dots \quad (99)$$

Similarly with the negative sequence network, starting at any point of zero potential within one of the machines and tracing one's way to F it will be found that

$$\overset{v}{e_2} = 0 - \Sigma \text{ voltage drops to the point F} \quad \ldots\ldots\ldots (100)$$

and for the zero sequence network

$$\overset{v}{e_0} = 0 - \Sigma \text{ voltage drops to the point F} \quad \ldots\ldots\ldots (101)$$

Adding these three voltages and equating to zero as indicated by equation (98):

$$\left. \begin{array}{l} 0 = \overset{v}{e}{'} - \Sigma \text{ voltage drops from } \overset{v}{e}{'} \text{ to F in} \\ \qquad \text{positive sequence network} \\[6pt] - \Sigma \text{ voltage drops from machines to F} \\ \qquad \text{in negative sequence network} \\[6pt] - \Sigma \text{ voltage drop from machines to F} \\ \qquad \text{in zero sequence network} \end{array} \right\} \ldots (102)$$

or

$$\left. \begin{array}{l} \overset{v}{e}{'} = \Sigma \text{ voltage drops from } \overset{v}{e}{'} \text{ to F in} \\ \qquad \text{positive sequence network} \\[6pt] + \Sigma \text{ voltage drops from machines to F} \\ \qquad \text{in negative sequence network} \\[6pt] + \Sigma \text{ voltage drops from machines to F} \\ \qquad \text{in zero sequence network} \end{array} \right\} \ldots (103)$$

If the potential of F_0 be made equal to zero, i.e., the neutral of the balanced system, it can be seen that equation (103), and consequently also equation (98) will be satisfied, and the connection of Fig. B-15 will completely determine the flow of current. The positive, negative, and zero sequence currents in any branch are equal to the actual currents flowing in the respective branches in the three networks.

The determination of the sequence and phase voltages and

currents at any point in the system will be discussed in connection with the application of the calculating board. This will insure a better picture of the correlation of the different networks and provide a better conception of the physical significance of the various quantities.

(2) *Double Fault to Ground.*—For this case assume both phases b and c faulted to ground simultaneously as shown in Fig. B-16 (a). The terminal conditions for this case are then

$$\left. \begin{array}{l} \overset{v}{e}_y = 0 \\[6pt] \overset{v}{e}_z = 0 \\[6pt] \overset{v}{I}_x = 0 \end{array} \right\} \quad \dots\dots\dots\dots\dots\dots\dots\dots\dots\dots\dots \quad (104)$$

From equations derived in the first article of this series

$$\left. \begin{array}{l} \overset{v}{e}_{x0} = 1/3(\overset{v}{e}_x + \overset{v}{e}_y + \overset{v}{e}_z) = 1/3(\overset{v}{e}_x) \\[6pt] \overset{v}{e}_{x1} = 1/3(\overset{v}{e}_x + a\,\overset{v}{e}_y + a^2\overset{v}{e}_z) = 1/3(\overset{v}{e}_x) \\[6pt] \overset{v}{e}_{x2} = 1/3(\overset{v}{e}_x + a^2\overset{v}{e}_y + a\,\overset{v}{e}_z) = 1/3(\overset{v}{e}_x) \end{array} \right\} \quad \dots\dots\dots \quad (105)$$

or

$$\overset{v}{e}_{x0} = \overset{v}{e}_{x1} = \overset{v}{e}_{x2} \quad \dots\dots\dots\dots\dots\dots\dots\dots\dots\dots \quad (106)$$

Also

$$\overset{v}{I}_x = \overset{v}{I}_{x0} + \overset{v}{I}_{x1} + \overset{v}{I}_{x2} = 0 \dots\dots\dots\dots\dots\dots\dots \quad (107)$$

Equations (106) and (107) define the conditions which must be fulfilled at the terminals for this case.

Connecting the three networks of Fig. B-13 as shown in Fig. B-16 it will be seen that the two above conditions are fulfilled; equation (107) is fulfilled at the junction of the three networks and equation (106) is fulfilled because the three voltages are measured across the same terminals and must therefore be equal to each other.

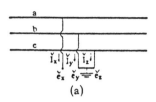

(a)

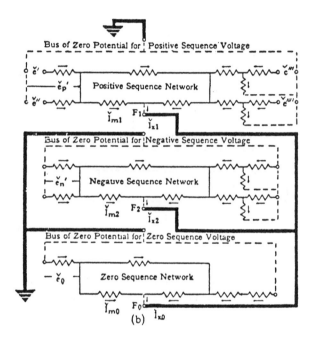

(b)

**Fig. B-16. The Three Sequence Networks for a
Double Fault to Ground**

(3) *Single-Phase Line-to-Line Fault.* — Assume the fault to
occur between phases b and c. The terminal conditions for
Fig. B-17 (a) are then

$$\left. \begin{array}{l} \overset{v}{I}_x = 0 \\[4pt] \overset{v}{I}_y = -\overset{v}{I}_z \\[4pt] \overset{v}{e}_y = \overset{v}{e}_z \end{array} \right\} \dots\dots\dots\dots\dots\dots\dots\dots\dots\dots\dots\dots (108)$$

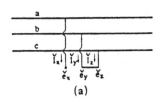

(a)

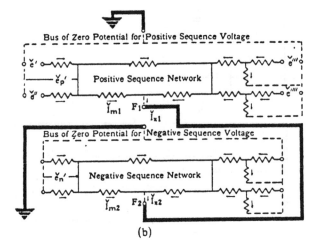

(b)

**Fig. B-17. The Three Sequence Networks for a
Single-Phase Line-to-Line Fault**

The sequence components are then

$$\overset{v}{I}_{X0} = 1/3(\overset{v}{I}_X + \overset{v}{I}_y + \overset{v}{I}_z) = 1/3(0 - \overset{v}{I}_z + \overset{v}{I}_z) = 0 \ \dots \ (109)$$

$$\overset{v}{I}_{X1} = 1/3(\overset{v}{I}_X + a\overset{v}{I}_y + a^2\overset{v}{I}_z) = 1/3(0 - a\overset{v}{I}_z + a^2\overset{v}{I}_z)$$

$$= \frac{-a + a^2}{3} \overset{v}{I}_z \dots\dots\dots\dots\dots\dots\dots\dots (110)$$

$$\overset{v}{I}_{X2} = 1/3(\overset{v}{I}_x + a^2\overset{v}{I}_y + a\overset{v}{I}_z) = 1/3(0 - a^2\overset{v}{I}_z + a\overset{v}{I}_z)$$

$$= \frac{+a - a^2}{3} \overset{v}{I}_z = -\overset{v}{I}_{X1} \dots\dots\dots\dots\dots\dots (111)$$

Since $\overset{v}{I}_{X0} = 0$ it follows that $\overset{v}{e}_{X0}$ must also be equal to zero. Furthermore,

$$\left.\begin{aligned}
\overset{v}{e}_{X1} &= 1/3(\overset{v}{e}_X + a\,\overset{v}{e}_y + a^2\,\overset{v}{e}_z) \\
\text{and } \overset{v}{e}_{X2} &= 1/3(\overset{v}{e}_X + a\,\overset{v}{e}_y + a\,\overset{v}{e}_z)
\end{aligned}\right\} \dots\dots\dots\dots (112)$$

Then, substituting in these equations the relations:

$$\overset{v}{e}_y = \overset{v}{e}_z$$

and $\overset{v}{e}_{X0} = 0$

$$\overset{v}{e}_{X1} = 1/3[\overset{v}{e}_X + (a + a^2)\,\overset{v}{e}_z] \dots\dots\dots\dots (113)$$

$$\overset{v}{e}_{X2} = 1/3[\overset{v}{e}_X + (a + a^2)\,\overset{v}{e}_z] \dots\dots\dots\dots (114)$$

$$= \overset{v}{e}_{X1} \dots\dots\dots\dots\dots\dots\dots\dots (115)$$

Because $\overset{v}{I}_{X0}$ and $\overset{v}{e}_{X0}$ are zero, the zero sequence network can be eliminated from consideration. The only conditions which must be fulfilled by the equivalent network are those expressed by equations (111) and (115). By connecting the positive and negative sequence networks as shown in Fig. B-17 it may be seen immediately that equation (111) is satisfied because $\overset{v}{I}_{X1}$ must always equal $-\overset{v}{I}_{X2}$. Equation (115) is satisfied because the equal voltages expressed thereby are measured from the same point and hence must be equal.

(4) *Three-Phase Faults.*—This case is the simplest and the one most familiar to persons calculating short-circuits for the application of circuit breakers and relays. The equivalent diagram reduces to that shown in Fig. B-18. Since the system remains balanced, the negative and zero sequence networks do not enter the problem.

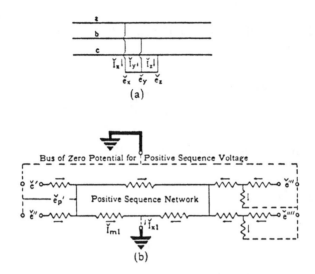

**Fig. B-18. The Three Sequence Networks for a
Three-Phase Short-Circuit**

APPLICATION TO CALCULATING BOARDS

The calculation of system faults by the methods just described
may be greatly facilitated by the use of calculating boards. By
means of an alternating-current board with resistors and reactors and several sources of e.m.f., the three complete sequence
networks representing the characteristics of the actual system
may be set up simultaneously. By appropriate connection of
these networks, in the manner just described, the different kinds
of faults can be represented. Regardless of the manner in which
the sequence networks are connected, the current in phase a at
any point can always be obtained by adding the currents of the
three networks at the corresponding points. Suppose it is desired to determine the current in the left-hand section of the
faulty line in Fig. B-13. The current in phase a is determined by
adding directly the current $\check{I}_{m1}$ from the positive-sequence network, $\check{I}_{m2}$ from the negative-sequence network, and $\check{I}_{m0}$ from

the zero-sequence network. This can be accomplished most conveniently by plugging into the three networks by means of current transformers and totalizing on the secondaries. For phases b and c, networks can be devised which shift the positive and negative sequence components 120 and 240 degrees; the shifted currents can then be added to the zero sequence current. In the absence of such a device the phase currents can be obtained by adding the quantities analytically after operating, by the unit vectors a and a^2, using the relations:

$$\overset{v}{I_b} = \overset{v}{I_{m0}} + a^2 \overset{v}{I_{m1}} + a \overset{v}{I_{m2}}$$
$$\overset{v}{I_c} = \overset{v}{I_{m0}} + a \overset{v}{I_{m1}} + a^2 \overset{v}{I_{m2}}$$

The sequence voltage for any point in a network appears only in the corresponding sequence network, so that the manner in which the sequence voltages are measured is independent of the particular way in which the networks are connected together for the representation of the different types of faults. In the positive sequence network the potential at any point is measured between that point and the point of zero or ground potential, indicated by the dotted line in the positive sequence network of Fig. B-13. Expressed differently, it is equal to one of the generated e.m.f.s minus the drop from the particular e.m.f. to the point considered. If the value of the particular e.m.f. be zero, the potential at the point will be the negative of the drop from the zero potential bus to the point. In the negative sequence network, since normally no generated e.m.f.s of negative sequence exist, the negative sequence potential at any point will be merely the negative of the drop to the point from the bus of zero negative potential. Similar considerations apply to the zero sequence network, the potential at any point being the negative of the drop to the point from the bus of zero potential for the zero sequence network as indicated by dotted lines in Fig. B-13. For example, the voltage of phase a on the left-hand bus is equal to the vector sum of the voltages

$\overset{v}{e}_p{}'$, $\overset{v}{e}_n{}'$, and $\overset{v}{e}_o{}'$, which are indicated in Fig. B-13. Note that this voltage is equal to the generated positive sequence voltage $\overset{v}{e}''$ minus the drops in the generator and transformer due to the positive, negative and zero sequence currents. Thus it may be observed that at the internal voltage of the machine all the drops are zero and the voltage of phase a is the generated positive sequence voltage. Also, for the case of a single conductor fault to ground as the point under consideration approaches the fault point, the voltage of phase a approaches zero. On the alternating-current calculating board the phase a voltage can be obtained by direct addition by means of potential transformers, and the voltages of phases b and c by phase shifting devices or analytically by means of the fundamental relations connecting phase and sequence voltages.

For short-circuit studies in which only the magnitudes of currents are required it is usually permissible to assume that the generated e.m.f.s are of the same magnitude and phase so that a single source may be used to represent the generated e.m.f. A further simplification results when it is found sufficiently accurate to assume all the impedances of the same phase angle. In this case the networks may be set up on a direct-current board, the elements of which are pure resistances.

When it is desired to obtain only the zero sequence currents or as they are sometimes called, residual or ground currents, the positive and negative sequence networks are required only for determining the magnitude of the fault current; the distribution of ground currents is determined of course only by the zero-sequence network.

Fault impedances may be introduced in the connections of the sequence networks for different types of faults and therefore, as will be shown in the next article, do not affect the method just described for measuring sequence and phase voltages and currents.

Appendix C

Review of Complex Numbers

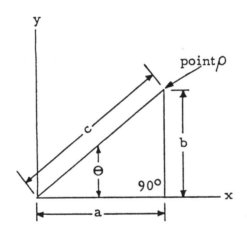

a = abscissa = c Cos Θ

b = ordinate = c Sin Θ

Cartesian Coordinates

Θ measured counterclockwise is positive

The specification of point ρ can be made in various forms as follows:

Rectangular	Complex	Exponential	Polar	Phasor	
a+jb	= c(CosΘ+jSinΘ) =	$ce^{j\Theta}$	= c$\underline{/\Theta°}$ =	$\dot{c}$	.... (1)
a−jb	= c(CosΘ−jSinΘ) =	$ce^{-j\Theta}$	= c$\underline{\angle-\Theta°}$ =	$\hat{c}$	.... (2)

a   is the real component
b   is the imaginary component
$c = \bar{c} = |c|$ is the modulus or absolute value (magnitude)
Θ is the argument or amplitude (relative phase
                 position)
$\dot{c}$ is a phasor
$\hat{c}$ is the conjugate of c

The absolute value of the phasor $\bar{c}$ or $|c| = \sqrt{a^2 + b^2}$ .... (3)

$$a = 1/2\,(\dot{c} + \hat{c}) \quad \text{by adding (1) and (2)} \ldots\ldots\ldots\ldots (4)$$

$$jb = 1/2\,(\dot{c} - \hat{c}) \quad \text{subtracting (1) and (2)} \ldots\ldots\ldots (5)$$

Multiplication Law: Absolute value of the product is the **product** of the absolute values of the components, and the argument is the **sum** of the component arguments. Thus:

$$\dot{E}\,\dot{I} = \bar{E} \times \bar{I} \;\big|\Theta_1 + \Theta_2 \ldots\ldots\ldots\ldots\ldots\ldots (6)$$

or $\quad \dot{E}\,\hat{I} = \bar{E}\,\epsilon^{j\Theta_1} \times \bar{I}\,\epsilon^{j\Theta_2} = \bar{E}\,\bar{I}\big|\Theta_1 - \Theta_2 \ldots\ldots\ldots\ldots (7)$

Division Law: This is the inverse of multiplication. Thus:

$$\frac{\dot{E}}{\dot{I}} = \frac{\bar{E}\epsilon^{j\Theta_1}}{\bar{I}\,\epsilon^{j\Theta_2}} = \frac{\bar{E}}{\bar{I}}\;\big|\Theta_1 - \Theta_2 \ldots\ldots\ldots\ldots\ldots\ldots (8)$$

Powers of Complex Numbers:

$$(\bar{I}\epsilon^{j\Theta})^n = (\bar{I})^n\,\epsilon^{jn\Theta} \ldots\ldots\ldots\ldots\ldots\ldots (9)$$

Thus: $\quad \bar{I}^2 = \bar{I}^2\,\epsilon^{j2\Theta}$

or $\quad \sqrt[n]{\bar{I}} = \sqrt[n]{\bar{I}\,\epsilon^{j\Theta}} = \sqrt[n]{\bar{I}}\,\epsilon^{j\frac{\Theta}{n}} \ldots\ldots\ldots\ldots (10)$

Phasor Times Its Conjugate:

$$\dot{I}\,\hat{I} = \bar{I}\,\epsilon^{j\Theta}\,\bar{I}\,\epsilon^{-j\Theta} = \bar{I}^2\,\epsilon^{j\,(\Theta - \Theta)} = \bar{I}^2 \ldots\ldots\ldots\ldots (11)$$

Appendix D

Transmission and Distribution Delivery Systems Efficiencies

Type of Circuit

1. Single-phase, 2-wire

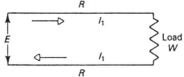

Conductor	Power Loss ($I^3 R$) in Conductor	Voltage Drop (IR)
I_1	$I_1^2 \times 2R = 2I_1^2R$	$I_1 \times 2R = 2I_1\,R$

2. Single-Phase, 3-wire

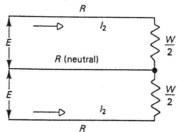

$$I_2 = \tfrac{1}{2}I_1 \qquad \begin{aligned} I_2^2 \times 2R &= (\tfrac{1}{2}I_1)^2 \times 2R \\ &= \tfrac{1}{4}(2I_1^2R) \end{aligned} \qquad \begin{aligned} I_2 \times R &= \tfrac{1}{2}I_1 \times R \\ &= \tfrac{1}{4}(2I_1R) \end{aligned}$$

<u>Type of Circuit</u>

3. Two-Phase, 4-wire

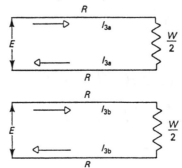

Conductor	Power Loss ($I^3 R$) in Conductor	Voltage Drop (IR)
$I_{3a} = \frac{1}{2}I_1$	$I_3{}^2 \times 2R \times 2$	$I_3 \times 2R = 2I_3 R$
$I_{3b} = \frac{1}{2}I_1$	$\quad = 4I_3{}^2 \times R$	$\quad = 2(\frac{1}{2}I_1)R$
	$\quad = 4(\frac{1}{2}I_1)^2$	$\quad = \frac{1}{2}(2I_1 R)$
	$\quad = \frac{1}{2}(2I_1{}^2 R)$	

4. Two-Phase, 3-wire

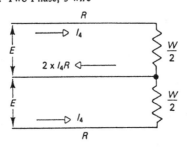

$I_4 = \frac{1}{2}I_1$	$I_4{}^2 = 2R + \left(\sqrt{2I_4}\right)^2 R$	$I_4 R + \sqrt{2I_4}R$
	$\quad = I_4{}^2(2R + 2R)$	$\quad = I_4 R + 1.41\ I_4 R$
	$\quad = 4I_4{}^2 R$	$\quad = 2.42\ I_4 R$
	$\quad = 4(\frac{1}{2}I_1)^2 R$	$\quad = 2.42(\frac{1}{2}I_1)R$
	$\quad = \frac{1}{2}(2I_1{}^2 R)$	$\quad = 1.21 \times I_1 R$
		$\quad = $ approx $\frac{1}{2}(2I_1)R$

5. Two-Phase, 5-wire

Since loads are balanced, no current will flow in the fifth or neutral wire. Hence, current, power loss, and voltage drop will be the same as for the Two-Phase, 4-wire system; refer to item 3 above.

Type of Circuit

6. Three-Phase, 3-wire — Y

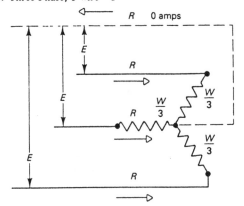

Conductor	Power Loss (I^3R) in Conductor	Voltage Drop (IR)
$I_6 = \dfrac{I_1}{3}$	$I_6{}^2 \times R \times 3$ $= \left(\dfrac{I_1}{3}\right)^2 \times R \times 3$ $= \dfrac{1}{3} I_1{}^2 R$ or $\dfrac{1}{6}(2I_1{}^2 R)$	$I_6 \times R = \dfrac{I_1}{3} = R$ $= \dfrac{1}{6}(2I_1 R)$

7. Three-Phase, 3-wire

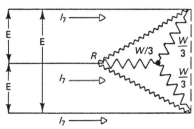

$$I_7 = \frac{I_1}{\sqrt{3}}$$

$I_7{}^2 \times R \times 3$

$= \left(\dfrac{1}{\sqrt{3}}\right)^2 \times R \times 3$

$= I_1{}^2 R$
or $\frac{1}{2}(2I_1{}^2 R)$

$I_7 \times R = \dfrac{I_1}{\sqrt{3}} R$

$= \dfrac{1}{2\sqrt{3}}(2I_1 R)$

*For comparison, this value

should be multiplied by $\sqrt{3}$
because line to neutral voltage
in this case is only $\dfrac{1}{\sqrt{3}}$ of E
assumed in case 1. Therefore,
comparative voltage drop
$= \frac{1}{2}(2I_1 R)$

Type of Circuit

---

8. Three-Phase, 4-wire — Y

Since loads are balanced, no current will flow in the fourth or neutral wire. Hence, current, power loss, and voltage drop will be the same as for Three-Phase, 3-wire — Y system; refer to item 6.

---

*Based on fixed, balanced loads at power factor of 1.0 and same wire size.

(Courtesy Long Island Lighting Co.)

Appendix E

Street Lighting— Constant Current Circuitry

Strictly speaking, street lighting is part of the distribution system. Some aspects, however, make it worthy of separate consideration. At one time these facilities were owned, installed, operated and maintained by the utility supplying electricity in the particular area. Municipalities now own the facilities, renting space on the utility pole, and employing independent contractors for maintenance. Paralleling these changes, the method of supply has also undergone change.

The early arc lights gave way to incandescent lamps connected in series, the 6.6 ampere rating being a carryover from the preceding arc lamps. The relatively long distance between street lights and the original limited development of the commercial distribution system made such systems practically ideal and universal, with the associated equipment and controls conveniently located in the area substation. With the growth of the distribution systems, street lights, connected in multiple and served from extensions of the commercial distribution system replaced the series system, employing a pilot wire controlled from the substation to operate the new street lights. Later, the pilot was controlled from photoelectric cells or time switches, and ultimately control of individual street lights from individual associated photoelectric relays proved economically feasible, and generally became the standard method of street light control.

A great number of series street lighting systems exist and will for some time to come; hence, they warrant further discussion. Diagrams illustrating a typical arrangement of a series street lighting circuit and the film disc cutout are shown in Figures E-1 and E-2.

Since the electric supply source is the primary voltage of a (practically) constant value, a device is needed to convert the supply to a constant current output for the series circuit. The constant current transformer has two coils, a primary that is a stationary constant voltage coil, and a movable constant current variable voltage coil. To maintain a constant current in the secondary coil, the voltage induced in it must

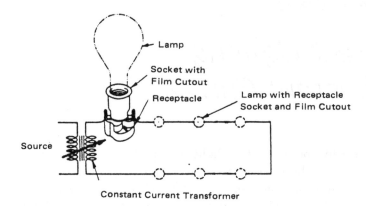

Figure E-1. Typical Series Street Lighting Circuit. *(Courtesy General Electric Co.)*

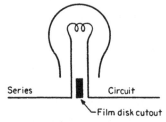

Figure E-2. Film Disc Cutout Schematic Diagram. *(Courtesy General Electric Co.)*

increase or decrease depending on the load, that is, the number of lamps in operation. The voltage variation is achieved by changing the position of the movable coil so that the coils operate closer or farther apart from one another. The position of the movable coil depends on three things:

1.   Force of gravity on the movable coil

2.   Force of attached weights countering the force of gravity on the movable coil

3.   Force of magnetic repulsion between the coils

The force of repulsion is due to the interaction of the magnetic fields set up in the coils. The magnitude of this force varies, depending on the amount of current flowing in the coils. A large current would exert a greater force tending to push the coils farther apart, than would a small current. As the coils are separated farther and farther apart, less and less of the magnetic field cuts through the secondary coil and hence less voltage is generated in it.

The counterweight changes or affects the force of gravity on the movable coil. By adding or subtracting weights from the counterweight, the balance can be regulated and the transformer made to operate at any desired current value over a limited range.

The current regulation of a constant current transformer is very accurate. This type of transformer cannot be overloaded because the secondary current will decrease if the load capacity of the unit is exceeded. The no-load condition corresponds to a short circuit on the secondary terminals and, hence, the movable coil approaches maximum separation. At full load, the coils attempt to separate further, but cannot because of the transformer construction. Beyond this point, any increase in load will result in a decrease in the secondary current. Operation of the constant current transformer is illustrated in Figure E-3, together with associated vector diagram.

Because of the constant current transformer characteristics, the I^2R loss remains constant for all load values. Stray losses increase with a decrease in load. The total loss increases with decrease in load and is maximum at no load. The operating temperature, therefore, is lower at full load than when operating at only a part of full-load rating.

Constant current transformers may have taps in the primary windings and sometimes on the secondary windings. They are constructed for both indoor or station operation as well as in tanks for outdoor and underground installation. They are rated in kVA.

PROTECTION

In a series circuit, should a film disc cutout fail to breakdown upon lamp failure or conductor break, the open circuit voltage established across the lamp terminals or open break in the conductor is proportional to the constant current or kW rating and is very high. For safety reasons, it is desirable that the circuit be deenergized, and this is done by means of an essentially high voltage relay incorporated as part of the remote control switch that deenergizes the primary supply to the constant current transformer.

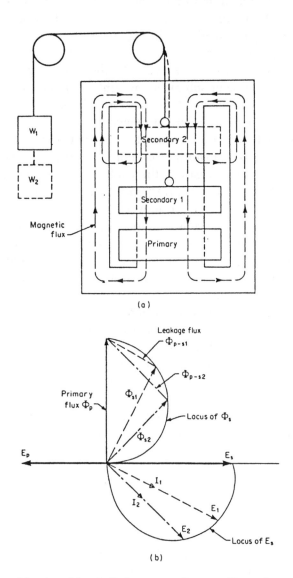

(a)

(b)

Figure E-3. (a) Movable-Coil Constant-Current Transformer. (b) Vector Diagram of Relation Between Primary and Secondary Voltage, Current, and Magnetic Flux. (Courtesy General Electric Co.)

Constant current transformers, both indoor and outdoor installations, are protected on both sides by surge arresters; those applied to the output or secondary side are based on open circuit voltage values. Where the lines may not be protected by shielding from conductors above them or by other structures, surge arresters may also be installed. Since the current in a series circuit is limited in value, grounding conductors are not subject to severe current flow. The importance of good grounds, however, should not be overlooked.

Film Disc Cutout

This is a disc that fits between the prongs of the socket that holds the lamp; they are the lamp terminals. Should the lamp fail, an open circuit is established between the terminals and a high voltage will appear. The film in the cutout is a special paper insulation that breaks down under this high voltage, causing a short circuit between the lamp terminals, which restores the continuity of the series circuit.

IL and SL Transformers

Because the voltage may be high at the socket of a series lamp. it is sometimes desirable to supply the lamp through a transformer that may reduce the voltage to safe values. The transformer is small and known as an "isolating" or "incandescent lamp" insulating transformer. The ratio of transformation is usually 1:1, although ratios may be different when supplying 15 or 20 ampere lamps. When this type transformer feeds several lamps in series, it is referred to as an SL transformer.

Series Circuit Control

This series street lighting circuit may be controlled directly at the substation by a time controlled switch, or may be controlled by pilot wire controlled by a switch operated by a time clock or a photoelectric relay; these are shown in Figures E-4 and E-5. Other pole type constant current transformers, usually of smaller ratings, may be cascaded from the circuit emanating from the substation, and may be controlled from the same control circuit.

LAMPS

In addition to the series and multiple type incandescent lamps, other types are also employed.

Luminous gas tubes, sometimes referred to as "neon" lights require voltages higher than secondary distribution voltages and are essentially constant current devices. The impedance of the tube increases as the length increases and inversely as the diameter of the tube. Autotransformers or two winding transformers, generally contained in the lamp fixture, are used to supply the necessary voltages.

Fluorescent lamps operate essentially in the same fashion as the gas tubes. The nominal secondary distribution supply voltage is raised to the lamp voltage by means of a "ballast" or small autotransformer incorporated in the lamp fixture.

High intensity lighting employs mercury, sodium or halogen lamps. In this type lamp, the material is vaporized into ionized gas that gives off colored light, different colors for the different materials. They are more efficient than incandescent lamps and give off less heat. Autotransformers are incorporated in each unit to supply the higher starting voltage (usually 600 volts) required.

These lamps may be controlled individually by individual relay or by a controlled circuit that supplies them. These lamps, sometimes referred to as "discharge type" lamps are usually designed for multiple circuit operation.

Diagrams of photoelectric controlled circuitry are shown in Figure E-4. Using solid-state circuitry, these are incorporated in small individual relays that usually control an individual lamp.

ACCESSORIES

Although not a part of this work, a general description of major accessories may be in order.

Street lights, without any accessories, in general emit light in all directions. If not directed, much of the light may not only be wasted, but may be a source of annoyance. To direct the light where it is needed and to employ it to maximum efficiency, its distribution is modified by the use of three general classes of accessories: reflectors, refractors, and diffusing glassware or other material. Lamps may be equipped with reflectors only, refractors only or diffusing globes only, or a combination of any of the three, Sometimes, the three are incorporated into one unit.

Street light fixtures may be mounted on poles with other distribution facilities, on separate poles of metal or concrete. They may be supplied from overhead or underground distribution lines. The fixtures come in a variety of shapes and sizes.

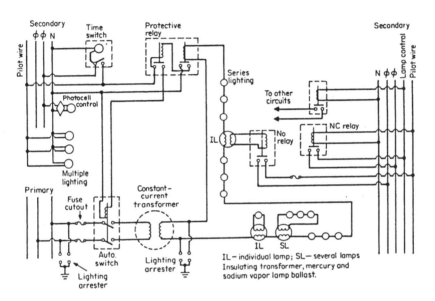

Figure 64. One-Line Diagram Series-Multiple and Multiple-Series Controlled Street Lighting. *(Courtesy General Electric Co.)*

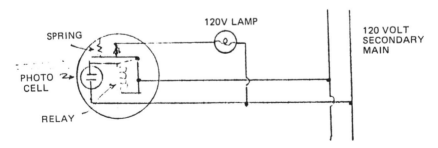

Figure E-5. Photoelectric Relay for Individual Lamp Control

Appendix 7

Economic Studies*

INTRODUCTION

Purpose

Economic studies are the means of evaluating the economic consequences of a particular proposal or of a number of alternate proposals for meeting a problem. Basic questions which continually face the management of any business are:

1. Will a venture be sufficiently profitable to justify the risk assumed in its undertaking?

2. Which of several ways of undertaking the venture will maximize the profits?

Scope

Economic studies may range from the extremely simple to the extremely complicated. In some cases, they may appear to be no more than the application of good common sense. The most important thing is the orientation which motivates a person to apply common sense or perform a more complicated evaluation of a situation.

Characteristics

No matter how complicated, economic studies all have certain definite characteristics.

1. Money to carry out every plan represents either:

---

a. Annual Expense- Obtained from operating revenue; or

b. Capital Expenditure -Obtained from financing, reinvested depreciation reserve, reinvested earnings. In general, capital costs represent the initial purchase price of installed plant; or

c. Both annual expense and capital expenditure.

2. Capital expenditures incur future annual expense.

3. The source usually available to a company to meet its annual expenses of operation, including taxes and obligations on its securities, is the revenue it receives from its consumers. Mathematically, therefore, the most economical of a number of plans (the one which will maximize profits) is the one which will require the minimum amount of additional revenue. A convenient way to conduct an economic study is to evaluate the effect of alternate proposals on the revenue requirements of the company.

4. Expenditures may take place (and thus affect revenue requirements) at different intervals over a period of time. The economic study must compare such expenditures on a consistent common basis.

5. The economics of alternate plans will generally be only one factor, although a major one, in the final selection of the most advantageous plan. Any differences in the nontangible items of comparison, however, must be recognized and considered with economic differentials among the several plans. The assignment of a value for the effect of inflation may be arbitrary and best omitted from the calculations and considered a judgmental factor in the final recommendations. (The effects of inflation at several rates are contained in Table F-1; other rates may be interpolated.) The differences in the various plans should be pointed out so that the phases of each alternative may be fully evaluated.

ANNUAL CHARGES

The overall revenue requirement of a project, or the cost of doing business, is the sum of annual charges for:

Table F-1. Inflation Factors (Compound Interest) $(I + I)^n$

Inflation rate—i in Percent

n	2	3	4	5	6	7	8	9	10
1	1.020	1.030	1.040	1.050	1.060	1.070	1.080	1.090	1.100
2	1.040	1.061	1.082	1.103	1.124	1.145	1.166	1.188	1.210
3	1.061	1.093	1.125	1.158	1.191	1.225	1.260	1.295	1.331
4	1.082	1.126	1.170	1.216	1.262	1.311	1.360	1.412	1.461
5	1.104	1.159	1.217	1.276	1.338	1.403	1.469	1.539	1.611
6	1.126	1.194	1.265	1.340	1.419	1.501	1.587	1.677	1.772
7	1.149	1.230	1.316	1.407	1.504	1.606	1.714	1.828	1.949
8	1.172	1.267	1.369	1.477	1.594	1.718	1.851	1.993	2.144
9	1.195	1.305	1.423	1.551	1.689	1.839	1.999	2.172	2.358
10	1.219	1.344	1.480	1.629	1.791	1.967	2.159	2.367	2.594
11	1.243	1.384	1.539	1.710	1.898	2.105	2.332	2.580	2.853
12	1.268	1.426	1.601	1.796	2.012	2.252	2.518	2.813	3.138
13	1.294	1.469	1.665	1.886	2.133	2.410	2.720	3.066	3.452
14	1.319	1.513	1.732	1.980	2.261	2.579	2.937	3.342	3.797
15	1.346	1.558	1.801	2.079	2.397	2.759	3.172	3.612	4.177
16	1.373	1.605	1.973	2.183	2.540	2.952	3.451	3.970	4.595
17	1.400	1.653	1.948	2.292	2.693	3.159	3.727	4.328	5.054
18	1.428	1.702	2.026	2.407	2.854	3.380	4.026	4.717	5.560
19	1.457	1.754	2.107	2.527	3.026	3.617	4.348	5.142	6.116
20	1.486	1.806	2.191	2.653	3.207	3.870	4.635	5.604	6.727
21	1.516	1.860	2.279	2.786	3.400	4.141	5.071	6.109	7.400
22	1.546	1.916	2.370	2.925	3.604	4.430	5.477	6.659	8.140
23	1.577	1.974	2.465	3.072	3.820	4.741	5.915	7.258	8.954
24	1.608	2.033	2.563	3.225	4.049	5.072	6.388	7.911	9.850
25	1.641	2.094	2.666	3.386	4.292	5.427	6.899	8.623	10.83
26	1.673	2.157	2.772	3.556	4.549	5.807	7.451	9.399	11.92
27	1.707	2.221	2.883	3.733	4.822	6.214	8.047	M.25	13.11
28	1.741	2.288	2.999	3.920	5.112	6.649	8.691	11.17	14.42
29	1.776	2.357	3.119	4.116	5.418	7.114	9.386	12.17	15.86
30	1.811	2.427	3.243	4.322	5.743	7.612	10.14	13.27	17.45
31	1.848	2.500	3.373	4.538	6.088	8.145	10.95	14.46	19.19
32	1.885	2.575	3.508	4.765	6.453	8.715	11.82	15.76	21.11
33	1.922	2.652	3.648	5.003	6.841	9.325	12.77	17.18	23.23
34	1.961	2.732	3.794	5.253	7.251	9.9 7 8	13.79	18.73	25.55
35	2.000	2.814	3.946	5.516	7.686	10.68	14.90	20.41	28.10

1.   Return on investment (stockholder, bondholder, etc.)
2.   Depreciation (sinking fund, etc.)
3.   Insurance expense
4.   Property tax expense
5.   Income tax expense
6.   Operating and maintenance expense
7.   Other taxes (e.g., on gross revenue)

The first five of these charges can usually, for convenience, be estimated as a percentage of original investment. The operating and maintenance charges should be separately estimated for each project. The tax (if any) on gross revenue must be calculated after all other charges are determined.

Return on Investment

A growing company must continually provide money for capital construction. In many cases, a large proportion of such funds are realized by sale of securities, bonds, debentures and stocks. These securities are purchased by people who believe that the future earnings of the company will provide a return on their investment commensurate with the hazards of the business and the nature of the security purchased. If the return provided is not sufficient to meet the expectations of investors when they analyze the risk involved, they will not invest in that firm. It is axiomatic then, that if a company is to be able to attract the necessary capital for continued expansion, it must maintain an adequate return on its invested capital.

Depreciation

The purpose of a depreciation allowance is to set aside a sufficient amount periodically (usually each year) to accumulate, over the life of the equipment, the original capital investment less net salvage.

There are a number of ways of taking account of depreciation; among the many types are two aptly named Straight Line Depreciation and Annuity Depreciation.

Straight Line Depreciation — The straight line method of calculating depreciation means that a fixed percentage is applied to surviving plant each year (usually monthly) to determine the accrual. The accrual rate is determined from the reciprocal of the average service life adjusted for salvage. This may be expressed by the equation:

$$D = \frac{1}{S}(1 - SAL)$$

where D = straight line depreciation rate
S = average service life
SAL = salvage ratio

Annuity Depreciation—It is also possible to express the annual charge for depreciation as an equivalent uniform annual charge. In cases where there is no salvage or dispersion (Iowa SQ dispersion), the annuity may be found in the future worth-to-annuity column in the compound interest table. This factor is determined from the equation:

$$AA = \frac{i}{(1 + i)^n - 1}$$

where AA annuity depreciation rate
i return as a percent of investment
n number of years

Dispersion is a factor to be considered in depreciable plant accounts. From actuarial studies, the nearest (Iowa) dispersion curve for each plant account has been previously determined. Thus to determine the annuity depreciation for a dispersed plant, the above equation is modified:

$$AA = \left(\frac{1}{\sum_{n-1}^{m} \frac{R_n}{(1 + i)^n}} - 1 \right) \times (1 - SAL)$$

where, in addition to above:
m = maximum or total life
R_n = mean annual survivor ratio
SAL = salvage ratio

The annuity depreciation factors for a dispersed plant have been calculated for every (Iowa) curve; please refer to Figure F-1. An elementary treatment of Depreciation, for illustrative purposes, is given in Table F-2.

There are many other ways of considering depreciation, and reference should be made to appropriate treatises on this subject.

Insurance

Unless specifically known, a value of 0. 1% of original investment generally is sufficient to be used for insurance.

The four major forms of insurance carried to provide protection against damage to property are:

1.   Fire insurance.

2.   Boiler and machinery insurance covering accidental damage to such objects.

3.   Coverage against damages due to motor vehicle collision, falling aircraft, storms, etc.

4.   Insurance for general liability in excess of some value (e.g., $50,000).

The premium expense of such insurance, expressed as a percentage of total investment, is usually very small, and the value of 0.1% as an average annual charge adequately covers premiums on the insurance carried. In any special case where items of insurance make up a substantial portion of operating expense, they should be considered separately in the estimation of operating expense.

Property Taxes

Taxes on property fall into two classes: special franchise or business taxes (applied to facilities on public property and to certain businesses); and, real estate taxes (applied to facilities on private property). Plant property classified as "Land Rights" (easements) or "Personal Property" such as tools, furniture and vehicles is not usually taxable. Depreciation is theoretically allowed on property, but in practice it is often not considered in computing taxes.

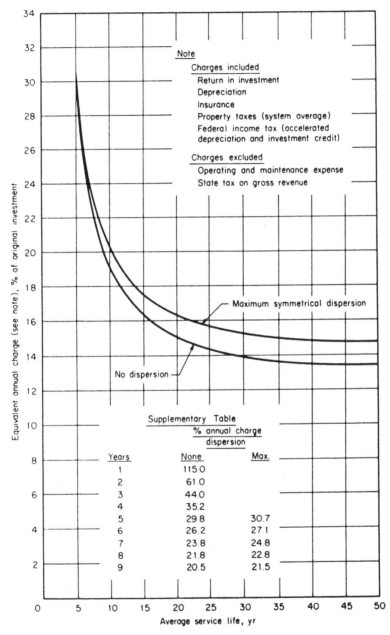

Figure F-1. Equivalent Annual Charges as a Percentage of Original Investment, Assuming No Salvage on Project and 7 Percent Return on Investment

Table F-2. Treatment of Depreciation
$1000 Capital Investment - 5 Year Life

Year	Investment at beginning	Depreciation	7% return	Total annual cost	Present worth Factor	Amount
A. Straight line depreciation						
1	$1,000	$200	$70	$270	0.935	$252
2	800	200	56	256	0.873	224
3	600	200	42	242	0.816	198
4	400	200	28	228	0.763	174
5	200	200	14	214	0.713	152
						1000
B. Very slow depreciation						
1	$1,0(0	0	70	70	0.935	65
2	1,000	0	70	70	0.873	61
3	1,000	0	70	70	0.816	57
4	1,000	0	70	70	0.763	54
5	1,000	1000	70	1070	0.713	763
						1000
C. Very fast depreciation						
1	$1,000	1000	70	1070	0.953	1000
2	0	0	0	0	0.873	0
3	0	0	0	0	0.816	0
4	0	0	0	0	0.763	0
5	0	0	0	0	0.713	0
						1000
D. Sinking fund depreciation (sinking fund factor from interest table)						
1	$1000	174	70	244	0.935	228
2		174	70	244	0.873	213
3		174	70	244	0.816	199
4		174	70	244	0.763	186
5		174	70	244	0.713	174
						1000

Federal Income Tax

Federal income taxes are levied on taxable income as defined in applicable laws. The relationship of taxable income to revenue and to return on investment is illustrated by Figure F-2 (a) and (b). Since rate of return on a project is the ratio of income available from the project to the net (depreciated) investment in the project, income tax must be calculated on the same basis of income or return.

Operating and Maintenance Expenses

Operating and maintenance expenses are constituent parts of the total annual charge. As a general rule, operating and maintenance expenses cannot be expressed as a percent of the plant or unit of property investment since they do not vary directly with the investment cost. Expenses must be specifically estimated based on the individual project and must include applicable loadings as well as direct charges. In the comparison of alternate plans, costs common to the plans in the same year may be eliminated since their difference will be zero. Large nonrecurring expenses must be evaluated on a present worth basis in the year of their occurrence.

Gross Earnings Tax

Some states (e.g., New York) levy taxes which are based on gross revenues. In evaluating alternate plans, the variation in this charge among plans will not affect the relative conclusions, and its consideration may be omitted unless total revenue requirements are desired.

BROAD ANNUAL CHARGE

For a complete study, it is necessary to evaluate the annual charges as discussed above applicable to a particular project. For many comparisons, enough accuracy can be obtained by using a more practical method employing broad annual charges. When the average service life exceeds 25 years, a broad annual charge of 15% of original investment may be used as a rough estimate of all charges exclusive of operating and maintenance expenses and gross earnings taxes. Figure F-1 shows the variation in total annual charge with service life and indicates the approximate basis for the 15% value.

This overall charge of 15% of original investment should not be

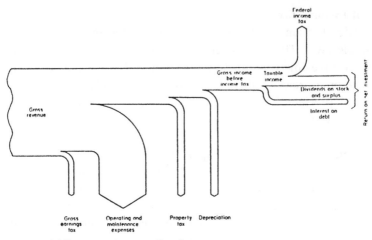

(a) Revenue and expense flow diagram.

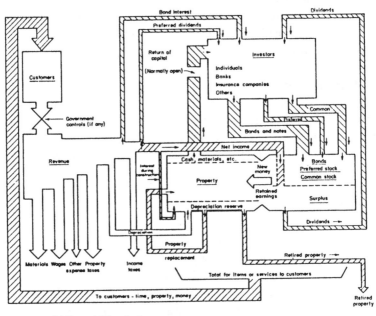

(b) Financial flowchart.

Figure F-2

applied to projects with a service life of less than 25 years.

The annual charges on projects with a service life of less than 25 years increase rapidly as service life shortens, as shown in Figure F-1. For such projects, a value determined from the upper curve of Figure F-1 for the service life of the particular project will provide a reasonable first approximation of the annual charge.

TIME VALUE OF MONEY

Earning Power

Money has earning power. A dollar today is worth more than a dollar a year from now because of this earning power available through investment. The precise value of today's dollar in the future will depend upon the rate of interest earned on the invested dollar. Thus, one dollar today, invested at a 7% interest rate, will be worth $1.07 one year in the future. Conversely, $1.07 available a year from now has a present worth of $1.00. By using this concept, that money has an increasing value over a period of time, any expenditure in the future may be expressed in its equivalent "present worth" today. This principle is used to convert expenditures made at varying times to an equivalent value at any one given instant.

Conversions

Such conversions may be made by converting values to:

1. Present worth-the value today.
2. Future worth-value at any specified time in the future.
3. Annuity-a uniform series of payments over a period of time.

The result of spending capital money is a series of annual charges extending over the service life of the property in which the capital is invested. Some of these annual charges will be uniform every year and may be considered an annuity. Other annual charges will vary from year to year resulting in a nonuniform series; these can be converted to a uniform series.

Conversion factors at 7% interest for all these manipulations are provided in Table F-3.

There are a number of different ways of developing the conversion factors. The convention used in Table F-3 is that annuity payments and future worth values are evaluated at the end of periods (years) and present worth values are evaluated at the beginning of periods. Developed in this way, Table F-3 is in its most directly usable form since all payments are assumed to be made at the end of a year (December 3 1) throughout this study.

EXAMPLES

Eight conversions cover all cases and are summarized and illustrated in the following examples and worth-time diagrams.*

F-1. Present worth to future worth (single amount at any date to single amount at any subsequent date)

You have just won $5,000, tax free. How much money will you have at the end of 10 years, if you invest it at 7% compounded annually?

Solution: See Figure F-3 (a). The $5,000 is a present worth, the value 10 years hence is a future worth. The future worth is obtained by multiplying the present worth by the conversion factor "Present Worth to Future Worth" for 10 years from Table F-3.

Future worth in 10 years = $5,000 × 1.967 = $9,835

EXAMPLE F-2. Future worth to present worth (single amount at any date to single amount at any previous date)

You have estimated that 10 years from now the unpaid mortgage on your house will be $9,835. How much money do you have to invest today at 7% interest to just accumulate $9,83 5 in 10 years?

Solution. The $9,835 is a future worth; the present worth of that amount is desired. From Table F-3, the conversion factor is 0.5083:

*Courtesy Long Island Lighting Co.

Table F-3. Compound Interest Table
i= 7%

	Lumpsum		Uniform annual series			
		Future				
	Present worth to future worth,	*worth to present worth,*	*Annuity to future worth*	*Future worth to annuity*	*Annuity to present worth*	*Present worth to annuity*
		1	$(1 + i)^n - 1$	i	$(1 + i)^n - 1$	$i(1 + i)^n$
n	$(1 + i)^n$	$\overline{(1 + i)^n}$	$\overline{i}$	$\overline{(1 + i)^n - 1}$	$\overline{i(1 + i)^n}$	$\overline{(1 + i)^n - 1}$
1	1.070	0.9346	1.000	1.00000	0.935	1.07000
2	1.145	0.8734	2.070	0.48309	1.808	0.55309
3	1.225	0.8163	3.215	0.31105	2.624	0.38105
4	1.311	0.7629	4.440	0.22523	3.387	0.29523
5	1.403	0.7130	5.751	0.17389	4.100	0.24389
6	1.501	0.6663	7.153	0.13980	4.767	0.20980
7	1.606	0.6227	8.654	0.11555	5.389	0.18555
8	1.718	0.5820	10.260	0.09747	5.971	0.16747
9	1.838	0.5439	11.978	0.083.19	6.515	0.15319
10	1.967	0.5083	13.816	0.07238	7.024	0.11238
11	2.105	0.4751	15.784	0.06336	7.499	0.13336
12	2.252	0.4440	17.888	0.05590	7.943	0.12590
13	2.410	0.4150	20.1.11	0.04965	8.358	0.11965
14	2.579	0.3878	22.550	0.04134	8.745	0.11434
15	2.759	0.3624	25.129	0.03979	9.108	0.10979
16	2.952	0.3387	27.888	0.03586	9.447	0.10586
17	3.159	0.3166	30.840	0.03243	9.763	0.10243
18	3.380	0.2959	33.999	0.02941	10.059	0.09941
19	3.617	0.2765	37.379	0.02675	10.336	0.09675
20	3.870	0.2584	40.995	0.02439	10.594	0.09139
21	4.141	0.2415	44.865	0.02229	10.836	0.09229
22	4.430	0.2257	49.006	0.02041	11.061	0.09041
23	4.741	0.2109	53.436	0.01871	11.272	0.08871
24	5.072	0.1971	58.177	0.01719	11.469	0.08719
25	5.427	0.1842	63.249	0.01581	11.654	0.08581
26	5.807	0.1722	68.676	0.01456	11.826	0.08456
27	6.214	0.1609	74.484	0.01343	11.987	0.08313
28	6.649	0.1504	80.698	0.01239	12.137	0.08239
29	7.114	0.1406	87.347	0.01145	12.278	0.08115
30	7.612	0.1314	94.461	0.01059	12.409	0.08059

Present worth = $9,835 × 0.5083 = $5,000

This is the reverse of Example F-1. The conversion factor for future worth to present worth is simply the reciprocal of the present worth to future worth factor. The worth-time-diagram is the same as for Example F-1.

EXAMPLE F-3. Annuity to future worth (annuity over any period to single amount at end of period)

You plan to save $500 of your earnings each year for the next 10 years. How much money will you have at the end of the 10th year if you invest your savings at 7% per year?

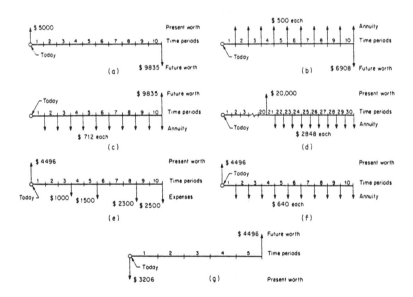

Figure F-3. (a) Illustrating Present Worth to Future Worth.
(b) Illustrating Annuity to Future Worth.
(c) Illustrating Future Worth to Annuity.
(d) Illustrating Present Worth to Annuity.
(e) Illustrating Nonuniform Expense.
(f) Illustrating Uniform Annual Charge.
(g) Illustrating Present Worth, Years Hence vs Today.

Solution. The $500 each year is an annuity since it is a uniform amount each year. You wish to know the future worth. From Table F-3, the annuity to future worth factor, 10 years, is 13.816:

Future worth of the annuity = $500 × 13.816 = $6,908

Note from Figure F-3 (b) that annuity payments are assumed to be made at the end of each time period. The conversion factor evaluates future worth at the same time that the last annuity payment is made.

EXAMPLE F-4. Future worth to annuity (single amount at any given date to annuity over any previous period ending at the given date)

If the unpaid mortgage on your house in 10 years will be $9,835, how much money do you have to invest annually at 7% interest to have just this amount on hand at the end of the 10th year?

Solution. See Figure F-3 (c). The $9,835 is a future worth; the uniform amount (annuity to set aside annually) is desired. From Table F-3, the future worth to annuity factor, 10 years, is 0.07238:

Annuity = $9,835 × 0.07328 = $712

EXAMPLE F-5. Present worth to annuity (single amount at any given date)

You hold an endowment type insurance policy which will pay you a lump sum of $20,000 when you reach age 65. If you invest this money at 7% interest, how much money can you withdraw from your account each year so that at the end of 10 years, there will be nothing left?

Solution. See Figure F-3 (d). The $20,000 can be considered the present worth at the end of the 10th year. From Table F-3, the present worth to annuity factor, 10 years, is 0. 1423 8:

Annuity which may be withdrawn for 10 years = S 20,000

$$\times\ 0.14238 = \$2,848$$

Note that direct use of the conversion factor assumes the first withdrawal to take place one period after the lump sum of $20,000 is received.

EXAMPLE F-6. Annuity to present worth (annuity over any period to single amount at start of the period)

> You have estimated that for the first 10 years after you retire you will require an annual income of $2,848. How much money must you have invested at 7% at age 65 to realize just this annual income?

Solution. The present worth of an annuity for 10 years is desired. From Table F-3, the annuity to present worth factor, 10 years, is 7.024:

$$\text{Present worth} = \$2,848 \times 7.024 = \$20,000$$

The worth-time diagram is the same as for Example F-5.

EXAMPLE F-7. Present worth nonuniform expenses to equivalent uniform annual charge

> The maintenance expenses for the next 10 years on a piece of equipment are estimated as follows:

Year	Amount
3	$1,000
5	1,500
8	2,300
10	2,500

> What is the present worth of these expenses? What is the uniform annual payment for 10 years equivalent to this nonuniform series? What does this mean?

Solution. See Figure F-3 (e). The expense amounts are future worths in the year indicated. The present worth is desired. Future worth to present worth factors from Table F-3:

Year	Amount	Factor future worth to present worth	Present worth
3	$1,000	0.8163	$ 816
5	1,500	0.7130	1,070
8	2,300	0.5820	1,339
10	2,500	0.5083	1,271
	Total		$4,496

The total present worth of these nonuniform series of expenses is $4,496.

The equivalent uniform annual series is obtained by applying the present worth to annuity factor for 10 years to the present worth. From Table F-3, present worth to annuity factor, 10 years, is 0. 1423 8:

$$\text{Equivalent uniform annual charge} = \$4,496$$

$$\times\ 0.14238 = \$640$$

(See Figure F-3 (f). This means that if you had $4,496 and invested it at 7%, you could withdraw the required amounts to meet exactly either the nonuniform series of expenses or pay out an equivalent amount of $640.

EXAMPLE F-8. Present worth some years hence to present worth today

Assume the expenses given in Example F-7 were to be associated with a piece of equipment to be installed 5 years from now. What is the present worth of the nonuniform expenses in that case?

Solution. See Figure F-3 (g). The present worth previously obtained was the present worth for the expenses incurred in the 10 years following installation of the project. This is a present worth 5 years from now. In terms of today's present worth, it is a future worth 5

years away. The present worth today is obtained simply by converting the future worth in 5 years to a present worth. From Table F-3, the future worth to present worth factor, 5 years, is 0.7130:

Present worth today = $4,496 × 0.7130 = $3,206

PROCEDURE FOR ECONOMIC STUDIES

The procedures for commencing an Economic Study may be laid out in a sequence of steps:

1. The facts concerning the different plans that could be used to meet the requirements of the problem should be set down. The plans should be made as comparable as possible.

2. The capital expenditures which will be incurred under each of the plans and the timing of these expenditures should be determined. The amounts and timing of operating and maintenance expenses must be estimated; allocations of cost to capital and expense must be adhered to.

3. A study period must be selected during which the revenue requirements incurred by the plans will be evaluated. In economic studies, it is seldom possible to find a study period which will precisely reflect the timing inherent in each of the plans under study. It will often be helpful to draw a diagram of the timing of capital and expense dollars for each of the plans in determining the study period. The study period chosen must be one determined on the basis of judgment. In every case, it must be sufficiently long to approximate the overall effects, over a long period of time, of the money reasonably to be spent for both capital and operating expenses.

4. The annual charges resulting from the capital expenditures in each phase must be calculated if broad annual charges cannot be applied. In considering alternate plans, items common to the several plans may be omitted from the calculations. The effect of temporary installations, salvage, and of the removal of equipment which

can be used elsewhere on the system must be taken into account.

5. When annual revenue requirements are nonuniform, the present worth of the revenue requirements for each plan must be calculated. The most economical plan will have the lowest present worth of revenue requirements. In the case where annual revenue requirements are uniform throughout the study period, the plan with the lowest annual requirements will be the most economical.

6. The comparison of the economic differences among the plans may be made on the dollar differences among the present worths of the revenue requirements. If percentage difference is considered, the dollar differences may be misleading as, in conducting the study, charges which are the same in the several plans are generally omitted; this will distort the base upon which a percentage difference is derived.

7. A recommendation of the most advantageous plan must be made. The plan with the minimum revenue requirements would be recommended from an economic point of view. Other considerations may indicate the recommendation of one of the other plans despite higher revenue requirements.

CONCLUSION

Economic studies constitute perhaps the most important ingredient in the implementation of a project. In sum, the consideration of any undertaking must answer satisfactorily three basic requirements or questions:

1. Why do it at all?
2. Why do it now?
3. Why do it this way?

The answers to these questions can, in large part, be supplied by the results of economic studies.

Appendix G

The Grid Coordinate System: Tying Maps to Computers*

Anthony J. Pansini, E.E., P.E.

INTRODUCTION

The grid coordinate system is the key that ties together two important tools, maps and computers. Maps are a necessity for the better operation of many enterprises, especially of utility systems. Their effectiveness can be increased many fold by adding to their information data contained in other files. Much of the latter data are now organized and stored in computer-oriented files-on punched cards and on magnetic tapes, drums, disks, and cells. Generally, these data can be retrieved almost instantly by CRTs (cathode ray tubes) or printouts. The link that makes the correlation of data contained on the maps and in the files practical is the grid coordinate system.

Essentially, the grid coordinate system divides any particular area served into any number of small areas in a grid pattern. By superimposing on a map a system of grid lines, and assigning numbers to each of the vertical and horizontal spacings, it is possible to define any of the small areas by two simple numbers. These numbers are not selected at random, but have some meaning. Like any graph, these two coordinates represent measurements from a reference point; in this respect they are similar to navigation's latitude and longitude measurements.

Further subdivision of the basic grid areas into a series of smaller grids is possible, each having a decimal relation with the previous one

---

(i.e., by dividing each horizontal and vertical space into tenths, each resultant area will be one-hundredth of the area considered). By using more detailed maps of smaller scale, it is possible to define smaller and smaller areas simply by carrying out the coordinate numbers to further decimals. For practical purposes, each of these grid areas should measure perhaps not more than 25 ft by 25 ft (preferably less, say 10 ft by 10 ft) and should be identified by a numeral of some 6 to 12 digits.

For example, by dividing by 10, an area of 1,000,000 ft by 1,000,000 ft (equivalent to some 190 miles square) can be divided into 10 smaller areas of 100,000 ft by 100,000 ft each, identified by two digits, one horizontal and one vertical. This smaller area can again be subdivided into 10 smaller areas of 10,000 ft by 10,000 ft each, identified by two more digits, or a total of four with reference to the basic 1,000,000-ft square area. Breaking down further into 1000- by 1000-ft squares and repeating the process allows these new grids to be identified by two more digits, or a total of six. Again dividing by 10 into units of 100 ft by 100 ft, and adding two more digits, produces a total of eight digits to identify this grid size. One more division produces grids of 10 ft by 10 ft and two more digits in the identifying number-for a total of 10 digits, not an excessive number to be handled for the grid size under consideration; see Figure G-1.

This process may be carried further where applications requiring smaller areas are desirable; however, each further breakdown not only reduces the accuracy of the measurements, but also adds to the number of digits, which soon becomes unmanageable. Experience indicates that a "comfortable" system should contain 10 digits or fewer for normal usages.

While the decimal relation has been mentioned, other relations can be used, such as sixths, eighths, etc., or combinations, such as eighths and tenths, and others.

Standard References

To give these numerals some actual physical or geographical significance, they may be tied in with existing local maps. U.S. Geological Survey maps, coast and geodetic survey maps, state plane coordinate systems, standard metropolitan statistical areas, or latitude and longitude bearings. They may also be tied in with maps independent of all of these.

While reference to state and federal government systems]ends

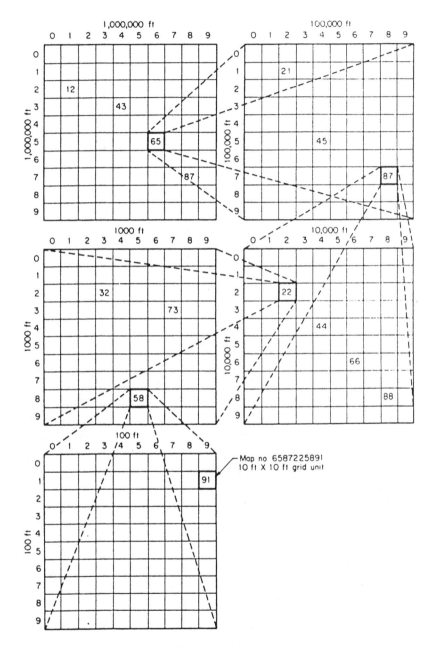

Figure G-1. Development of Grid Coordinate System

some geographical significance, it produces identifying grid numbers with several additional digits. It is not necessary for any grid coordinate system to have this reference to a government system, but if it is desired, it is a relatively simple procedure to develop a computerized look-up program that can translate such coordinates.

Basic grid coordinate maps may be developed from the conversion of existing maps to a usable scale, if such maps are reasonably accurate and complete, both as to their geography and content. They may also be developed from exact land surveys, from aerial surveys, or from combinations of all of these.

Excellent maps are also available for most of the country. U.S. Geological Survey maps show latitude and longitude lines every few miles; they also show numerous triangulation stations with the latitude and longitude for each station determined to an extreme degree of accuracy. Further, detail maps are available for practically every city and township, showing streets, houses, and lots. Despite the fine degree of accuracy of these maps, minor inaccuracies and discrepancies are bound to occur.

Earth's Curvature. Errors occur in mapping the earth's curved surface on a flat map; see Figure G-2. For example, in the case of the approximate 190-mi square mentioned previously, in the continental United States, the error introduced by this curvature, measuring from the center (95 mi in the longitudinal, or north-south direction) would probably not exceed 2 percent, a tolerable error. These errors need not be of great import, except in establishing match lines between maps. No gaps or overlaps should appear between adjacent maps, or between property or lot lines within a map. Tolerances of a few percent ordinarily are acceptable.

GRID COORDINATE MAPS

A grid coordinate map system should meet the following requirements:

1. It should include a simple and easily understood system of numerals for locating the data under consideration (numerals only; the x and y coordinates).

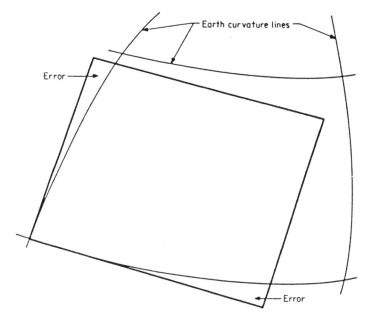

Earth curvature lines

Error

Error

Figure G-2. Error Introduced in Grid Coordinate System By Earth Curvature

2. The grid areas should be small enough to be consistent with the purposes for which they are to be used (25 ft or less).

3. The number of digits in the grid number should be held to a practical number so as not to become cumbersome and unwieldy (normally not more than about 10).

4. It should be designed to allow for expansion so that it will not have to be radically revised if unforeseeable expansion should occur.

5. It must provide reasonable accuracy (tolerances of a few percent), and it may or may not be tied in with some government or other established coordinate system.

6. Map sizes should be manageable (say, 24- or 36-in. square).

7. Maps of different scales should be included in the system to accommodate different kinds of data (for circuit data, say 1000 or 500 ft; for details of facilities, say, 100 ft for overhead and 50 or 25 ft for underground).

8. A key map must show the entire grid area.

9. Optional is a grid atlas showing street locations with grid overlay.

In attempting to design a grid coordinate system for a very large area, it may be difficult to meet these requirements. In such instances it may be desirable and practical to divide the entire area into two or more convenient districts, establishing a separate grid coordinate system in each district and tying the separate systems together with match lines at the borders. A prefix letter or number may be used to identify districts, though this may not prove necessary in actual operation.

The size of individual maps should be large enough to encompass an area suitable for the purpose but small enough not to be unwieldy; sizes 24- or 26-in. square have proven practical. Maps of different scales are used for different purposes; for example, a 50- or 100-ft scale is used for dense or crowded areas; 300- or 500-ft for less dense or rural areas; and 500- or 1000-ft or even larger for district or overall area viewing. The series of scales used should be such that the larger-scale maps fit into those of smaller scale completely and evenly. Match lines of each sheet should fall on corresponding match lines of adjacent sheets.

The grid pattern applicable to each of the several scale maps may be printed directly on each map as light background lines, perhaps even in a different color, or printed on the back of the maps when they are reproduced. Alternately, a grid overlay can be applied to each map to be used when it is necessary to determine a grid coordinate for an item on the map. The actual grid number need not be printed on every item on every map unless desired. Such numbers assigned to key locations on each map normally suffice; the others may be determined from the grid background or overlay. To maintain their permanence and to minimize distortion from expansion and contraction because of changing humidity and temperature, the maps should be printed on a material such as Mylar, a translucent polyesterbase plastic film; this is especially true of the base maps from which others of different scales and purposes are derived.

As much as practical, the data on the maps should be uncluttered and as legible as possible. It may be desirable in some instances to provide two or more maps (of the same scale) for several purposes; marks for coordinating these several maps, should it be necessary, may be included on each of the maps.

Maps may be further uncluttered by deliberately removing as much of the information on them as appears desirable and practical and consigning such information to files readily accessible by computer. In many instances, this information already is included or duplicated in such files, but may need to be labeled with the appropriate grid coordinate number. The use of CRTs and printouts makes this information available at will.

Application of Grid Coordinate Numbers

The grid coordinate number may be applied to each item of information contained in the computer-operated files by location. This may be done in several ways: manually, by machine, or a combination of the two.

The manual method is to superimpose or overlay the grid pattern on existing maps and manually assign numbers to each item to be processed. As mentioned previously, the grid pattern may be transferred to the master or original maps, and reproduced (or microfilmed) on the maps for the user; here numbers can be assigned directly from the map.

The machine method of grid number assignment employs an electronic scanning device called a *digitizer*. This machine includes a drafting table for map display and a cursor or pointer. The postal address and other fixed data are inserted on a punch card. For a particular map, the grid numbers of the map are set on the digitizer console. The digitizer assigns the x and y coordinates when the cursor is placed on a selected point and activated. These data are fed to a keypunch, which produces a punched card. In this method of producing the grid coordinate numbers, the digitizer enables additional refinement to be achieved, producing additional decimal numbers for the x and y coordinates. Hence, the ultimate grid area that can be measured can be one-tenth or one-hundredth, etc., of the basic unit area (1 by 1 ft, or 0. 1 ft by 0. 1 ft, for a 10- by 10-ft base area).

When numbers are assigned manually from maps, this degree of accuracy is not possible, nor is it necessary if the principal purpose of the grid coordinate system is to identify an item rather than a precise point. While the actual accuracy of such additional digits can be questioned, they provide a method of further subdividing a map for closer location of an item, but more important, they make possible a system of automatic mapping using the computer.

The grid coordinate number corresponding to the location of an item in question is added to its record and now becomes its computer *address*. In assigning these numbers to existing files, the digitizer generally can property identify the location from a suitable map. As a practical matter, however, there will be some locations or descriptions that cannot be identified using the digitizer, and these may require manual processing and actual checking in the field; fortunately these usually constitute only a small percentage of the total records.

In the maintenance of such files, the grid coordinate numbers associated with changes in, or with the introduction of new items into, the records can be assigned manually by the originator of the record.

COORDINATE DATA HANDLING

As implied earlier, the grid coordinate system provides an easy and simple but, more important, a very rapid means of obtaining data from files through the use of the computer. In some respects, it assigns addresses to data in the same way as the ZIP code system in use by the postal service. The manner in which the grid number may be used is illustrated in the following examples; for convenience they refer to electric utility systems, although obviously they apply equally well to other endeavors employing maps and records.

Data contained on maps and records generally apply to the consumers served and the facilities installed to serve them. While maps depict (by area) the geographic and functional (electrical) interrelationship between these several components, the records supply a continuing history (by location) of each component item (consumers and facilities).

In the case of consumers, such data may include, in addition to the grid coordinate number, the name and post office address. Also a history of electric consumption (and demand where applicable), billing, and other pertinent data over a continuing period, usually 18 or 24 months. There may also be data on the consumer's major appliances; also the data and work order number of original connection and subsequent changes. The grid coordinate number of the transformer from which the consumer is supplied is included, as well as that for the pole or underground facility from which the service to the consumer is taken. Sometimes interruption data may be included. Other data may include telephone number, tax district, access details, hazards (including animals), dates of connection or reconnection, insurance claims, casements, meter data, meter reading route, test data, credit rating, and other pertinent information. Only a small portion of these data are shown on maps, usually in the form of symbols or code letters and numerals. In the case of facilities, such data may include, in addition to the grid coordinate number, location information, size and kind of facility (e.g., pole, wire, transformer, etc.), date installed or changed, repairs or replacements made (including reason therefor, usually coded), original cost, work order numbers, crew or personnel doing work, construction standard reference, accident reports, insurance claims, operating record, test data, tax district, and other pertinent information. Similarly, only a small portion of these data are shown on maps, usually in the form of symbols or code letters and numerals.

Data from other sources also may be filed by grid number for correlation with consumer and facility information for a variety of purposes. Such data may include government census data; police records of crime, accidents, and vandalism; fire and health records; pollution measurements; public planning; construction and rehabilitation plans; zoning restrictions; rights-of-way and easement locations; legal data; plat and survey data; tax district; and much other information that may affect or be useful in carrying out utility operations.

Obviously, all these data, whether pertaining to the consumer or to the utility's facilities, are not necessarily contained on one map or in one record only; indeed, there may be several maps and records involved, each containing certain amounts of specialized or functionally related data. All, however, may be correlated through the grid coordinate system.

Data Retrieval

Data contained in the files may be retrieved by means of the computer and may be presented visually by means of CRTs for one-time instant use, or by printouts and automatic plotting for repeated use over time. Data presented may be the exact original data as contained in one or more files, or extracted data obtained as a result of correlating data residing in one or more files, or a combination of both; such extracted data may or may not be retained in separate files for future use.

These data may be retrieved for an individual consumer or an individual item of plant facilities, or may be other data for a particular area, small or large. The various specific purposes determine what data are to be retrieved and how they are to be presented. They also determine the programs and equipment required. Data thus retrieved then are used with data contained on the map to help in forming the decisions required. The decisions may include new data that can be reentered in the files as updating material, that can be plotted or printed for exhibit purposes, or that can be reentered on maps for updating or expanding the material thereon; all of these may be done by means of the computer.

The grid coordinate number is applied to utility facilities for case of location and positive identification in the field. In the case of electric utilities, these may include services, meters, poles, towers, manholes, pull boxes, transformers, transformer enclosures, switches, disconnects, fuses, lightning arresters, capacitors, regulators, boosters, streetlights, air pollution analyzers, and other equipment and apparatus; also the location of laterals on transmission and distribution circuits.

OTHER APPLICATIONS

Similarly, for gas utilities, the applications of grid coordinate numbers may include mains, services, meters, regulators, valves, sumps, test pits, and other equipment; also the location of boosters, laterals, and nodes on the gas systems. For water systems, they may include mains, services, meters, valves, dams, weirs, pumps, irrigation channels, and other facilities. For telephone and telegraph communication systems, including CATV circuits, they may include mains, services, terminals, repeaters, microwave reflectors, and other items including poles, manholes, and special items.

Grid coordinate numbers also may find application in many other lines of endeavor: highway systems, railway systems, oil fields, social surveys (police, health, income, population distribution, etc.), market surveys (banks and industries), municipal planning and land use studies, nonclassical archeology, geophysical studies, and others where such means of location identification may prove practical.

The use of grid coordinates facilitates positive identification in the field; the numbers are posted systematically on facilities, such as streetlight or traffic standards, poles, and structures, and at corners or other prominent locations.

An atlas, consisting of a grid overlay on a geographical map, aids the field forces in locating consumers and plant facilities and provides a common basis for communication between office and field operating personnel.

The grid pattern permits the classical manipulation of data by individual grid sections or areas comprising several grid sections. In addition to the sample presentation of such data by means of CRT displays and typed printouts, data may be presented in the form of plotting in various graphical forms, in patterns indicating the distribution of data, the density of particular data, the accumulation of data within fixed boundaries, the determination of area boundaries for predetermined data content (the analysis of data within a given polygon), the calculation of lengths and distances between grid locations, and the mapping of facilities in acceptable detail-and all of these operations may be performed automatically by means of the computer.

Further, summaries and analyses employing the grid coordinate system may be more readily made and are susceptible to combination and consolidation, resulting in perhaps fewer and more comprehensive reports (eliminating the duplication of much needless data and the presentation of more complete and meaningful conclusions in one place).

In all of the foregoing discussion, the point must be made that all of the handling of data using the grid system may also be accomplished without the use of the grid system. It is apparent, however, that this latter method will in the vast majority of cases employ more effort in terms of work hours and will be more time-consuming, so as to render many applications impractical, even though their desirability may be great; in short, the grid coordinate system enhances the economics of data handling.

ECONOMICS

It is not to be denied that the introduction of the grid coordinate system will impose additional cost to the maps and records function. It is also evident that these costs will be offset by the decreased personnel requirements in the processing of data derived from the maps and records, especially when the computer may be made to take up a large part of this burden. Moreover, more refinement and a wider scope in processing of data are attainable.

The cost of implementing a grid coordinate system can be evaluated fairly accurately. Many factors will influence the final determination; these include the area of the system involved, the number of consumers and facilities, the condition of the basic and auxiliary maps and records, the number and scope of the applications desired, the extent of automation, and many other factors. A very approximate estimate may average perhaps about one day's revenue per consumer. Practical considerations associated with implementation may well dictate a period of several years, perhaps 5 years or even more, over which the expenditure will have to be made to accomplish the desired goals.

The offsetting savings from the introduction of a grid coordinate system, including those derived from the additional Worth of the wider utilization, are difficult to pinpoint. It should be observed that while it is probable that a single application will not justify the adoption of the grid coordinate system, except in some unusual or special set of circumstances, it is also probable that the multiplicity of practical applications indicated will justify the relatively modest expenditure necessary for the conversion of present maps and records to the grid coordinate system.

The personnel requirements necessary to implement a grid coordinate system over a reasonable (short-term) period of time must be viewed together with the overall probable lessened longer-term in-house requirements. Since such a conversion is a one-time operation, it recommends itself admirably to the classical use of contractors having the special skills and experience. Further, such outside services are not apt to be diverted by crisis incidents prevalent in many enterprises.

One final observation. With the national consensus apparently pointing to an ultimate metric system for the United States to conform with world standards, the adoption of a grid coordinate system provides an excellent opportunity for its introduction with a minimum of conversion effort.

With the advent of the computer, it was inevitable that the grid coordinate system should be developed to provide a simple means of addressing the computer. The grid number provides the link between the map and the vast amount of data managed by the computer. This happy marriage of two powerful tools results not only in better operations but in improved economy as well. It is a must in the modernization of operations in many enterprises and especially in utility systems.

Appendix H

United States and Metric Relationships

U.S. To Metric		Metric To U.S.	
Length			
1 inch	= 25.4 mm	1 millimetre	= 0.03937 inch
1 inch	= 2.54 cm	1 centimeter	= 0.3937 inch
1 inch	= 0.0254 m	1 metre	= 39.37 inch
1 foot	= 0.3048 m	1 metre	= 3.2808 feet
1 yard	= 0.9144 m	1 metre	= 1.094 yard
1 mile	= 1.609 km	1 kilometre	= 0.6214 mile
Surface			
1 inch^2	= 645.2 mm^2	1 millimetre^2	= 0.00155 inch^2
1 inch^2	= 6.452 cm^2	1 centimetre^2	= 0.155 inch^2
1 foot^2	= 0.0929 m^2	1 metre^2	= 10.764 foot^2
1 yard^2	= 0.8361 m^2	1 metre^2	= 1.196 yard^2
1 acre	= 0.4047 hectare	1 hectare	= 2.471 acres
1 mile^2	= 258.99 hectare	1 hectare	= 0.00386 mi^2
1 mile^2	= 2.59 km^2	1 kilometre^2	= 0.3861 mile^2
Volume			
1 inch^3	= 16.39 cm^3	1 centimetre^3	= 0.061 inch^3
1 foot^3	= 0.0283 m^3	1 metre^3	= 35.314 foot^3
1 yard^3	= 0.7645 m^3	1 metre^3	= 1.308 yard^3
1 foot^3	= 28.32 litres	1 litre	= 0.0353 foot^3
1 inch^3	= 0.0164 litre	1 litre	= 61.023 inch^3
1 quart	= 0.9463 litre	1 litre	= 1.0567 quarts
1 gallon	= 3.7854 litres	1 litre	= 0.2642 gallons
1 gallon	= 0.0038 m^3	1 metre^3	= 264.17 gallons
Weight			
1 ounce	= 28.35 grams	1 gram	= 0.0353 ounce
1 pound	= 0.4536 kg	1 kilogram	= 2.2046 lb*
1 net ton	= 0.9072 T (metric)	1 Ton (metric)	= 1.1023 net tons**

*Avoirdupois
**1 ton = 2000 lb

Compound units			
1 lb/ft	= 1.4882 kg/m	1 kilogram/metre	= 0.6720 lb/ft
1 lb/in^2	= 0.0703 kg/m^2	1 kg/cm^2	= 14.223 lb/in^2

U.S. to Metric	Metric to U.S.

Compound units (continued)

1 lb/ft²	= 4.8825 kg/m²	1 kg/m²	= 0.2048 lb/ft²
1 lb/ft³	= 16.0192 kg/m³	1 kg/m³	= 0.0624 lb/ft³
1 ft-lb	= 0.1383 kg-m	1kg-m	= 7.233 ft-lbs
1 hp	= 0.746 kW	1 kW	= 1.340 hp
1 ft-lb/in²	= 0.0215 kg-m/cm²	1 kg-cm/m²	= 46.58 ft-lb/in²

Temperature

1 degree F	= 5/9 degree C	1 degree C	= 9/5 degree F
Temp °F	$= \dfrac{9}{5}\,°C + 32$	Temp °C	$= \dfrac{5}{9}\,(°F - 32)$

Index